OFFICIAL SQA PAST PAPERS

WITH ANSWERS

D1471592

STANDARD GRADE | GENERAL | CREDIT

GEOGRAPHY
2008-2012

First exam published in 2008.
Published by Bright Red Publishing Ltd, 6 Stafford Street, Edinburgh EH3 7AU
tel: 0131 220 5804 fax: 0131 220 6710 info@brightredpublishing.co.uk www.brightredpublishing.co.uk

ISBN 978-1-84948-247-9

A CIP Catalogue record for this book is available from the British Library.

Bright Red Publishing is grateful to the copyright holders, as credited on the final page of the Question Section, for permission to use their material. Every effort has been made to trace the copyright holders and to obtain their permission for the use of copyright material. Bright Red Publishing will be happy to receive information allowing us to rectify any error or omission in future editions.

[BLANK PAGE]

FOR OFFICIAL USE

G

KU   ES

Total Marks

1260/403

NATIONAL
QUALIFICATIONS
2008

FRIDAY, 9 MAY
10.25 AM–11.50 AM

GEOGRAPHY
STANDARD GRADE
General Level

Fill in these boxes and read what is printed below.

Full name of centre

Town

Forename(s)

Surname

Date of birth
Day Month Year   Scottish candidate number   Number of seat

1 Read the whole of each question carefully before you answer it.

2 Write in the spaces provided.

3 Where boxes like this ☐ are provided, put a tick ✓ in the box beside the answer you think is correct.

4 Try all the questions.

5 Do not give up the first time you get stuck: you may be able to answer later questions.

6 Extra paper may be obtained from the invigilator, if required.

7 Before leaving the examination room you must give this book to the invigilator. If you do not, you may lose all the marks for this paper.

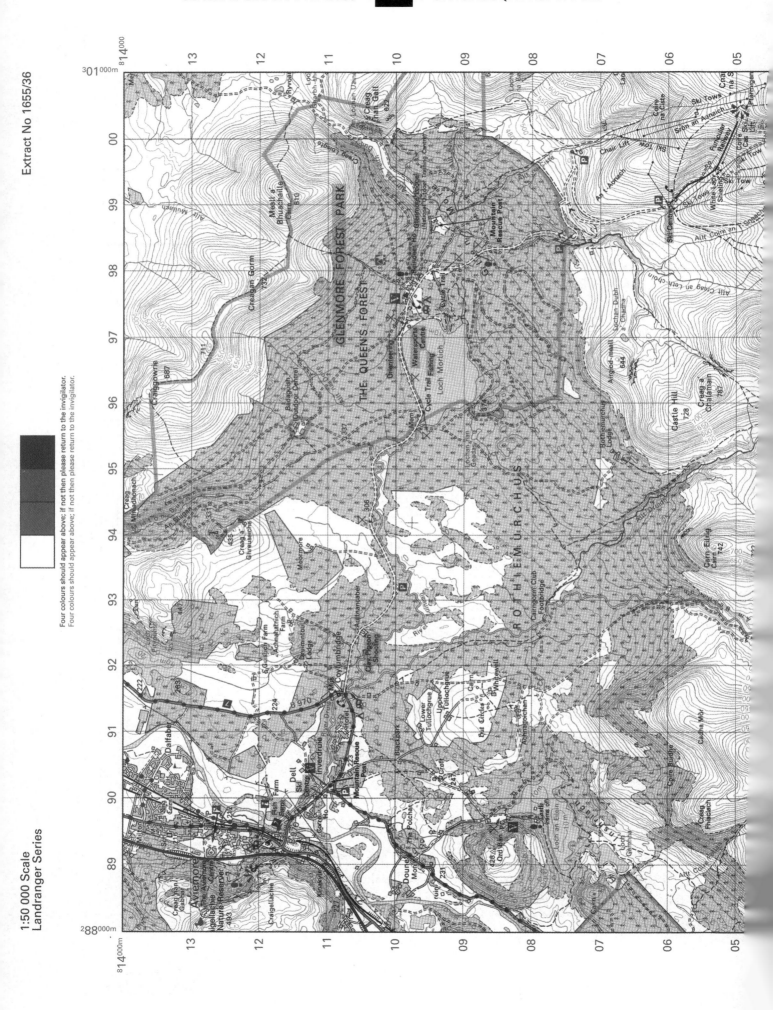

Extract No 1655/36

1:50 000 Scale
Landranger Series

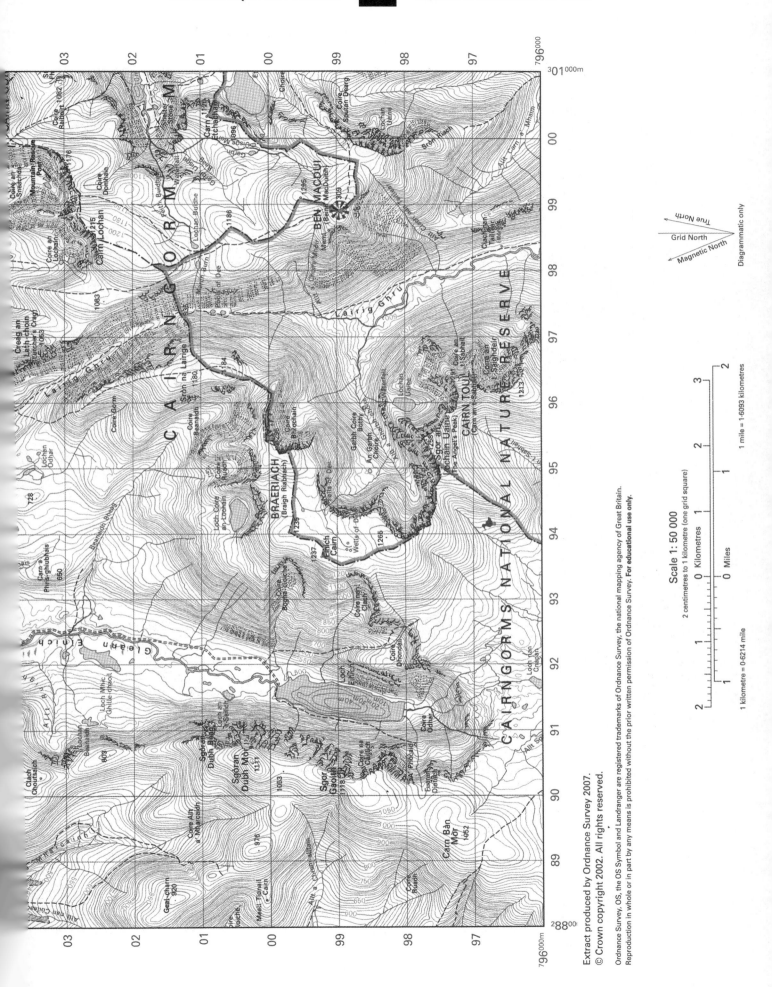

Scale 1 : 50 000

2 centimetres to 1 kilometre (one grid square)

1 kilometre = 0·6214 mile

1 mile = 1·6093 kilometres

True North
Grid North
Magnetic North

Diagrammatic only

Extract produced by Ordnance Survey 2007.
© Crown copyright 2002. All rights reserved.

1.

Reference Diagram Q1A: The Aviemore Area

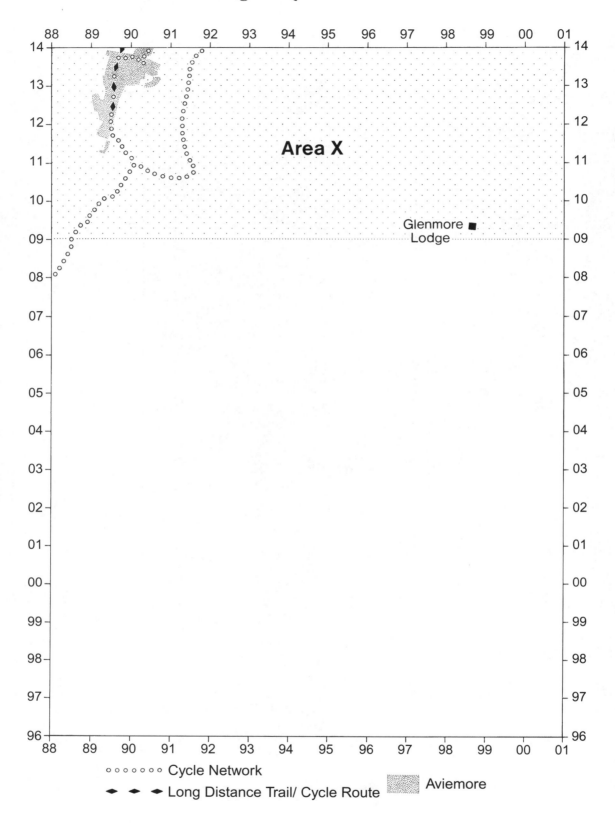

○ ○ ○ ○ ○ ○ ○ ○ ○ Cycle Network

◆ ◆ ◆ Long Distance Trail/ Cycle Route

Aviemore

Marks

1. (continued)

Look at the Ordnance Survey Map Extract (No 1655/36) of the Aviemore area and Reference Diagram Q1A on Page two.

(*a*)   Using the map extract, match the glaciated features A, B, C and D to the correct name/grid reference in the table below.

A   Ribbon Lake   B   Corrie   C   U shaped Valley   D   Pyramidal Peak

Grid Reference/Name of Feature	Letter
9597 Angel's Peak	
9798 Lairig Ghru	
9400 Loch Coire an Lochain	
9198 Loch Einich	

3

(*b*)   **Explain** how a pyramidal peak was formed.

You may use diagram(s) to illustrate your answer.

_____

_____

_____

_____

_____

_____

_____

_____

_____

_____

_____

3

Marks

1. (continued)

(c)   Find Area X on the map extract and on Reference Diagram Q1A.

In what ways has the physical landscape in Area X both encouraged **and** limited settlement growth?

_____

_____

_____

_____

_____

4

(d)   Glenmore Lodge (9809) is a National Outdoor Training Centre*.

(*This type of centre provides instruction in outdoor activities.)

How good is this location for the centre?

Give reasons for your answer.

_____

_____

_____

_____

_____

4

Marks

1. **(continued)**

    Reference Diagram Q1B: Selected Aims of National Parks

    - *Preserve beauty of countryside*
    - *Conserve local wildlife*
    - *Provide good access and facilities for public enjoyment*
    - *Maintain farming*

    (*e*)　This area is part of the Cairngorm National Park.

    　　Explain how land uses shown on the OS map **and** outdoor activities in this area might be in conflict with the aims shown in Reference Diagram Q1B.

    _____

    _____

    _____

    _____

    _____　**4**

    (*f*)　There is a Long Distance Trail and Cycle Network shown on both the map extract and Reference Diagram Q1A.

    　　What techniques could be used to gather information on the impact of walkers and cyclists on the local area?

    　　Give reasons for your choices.

    _____

    _____

    _____

    _____

    _____

    _____　**4**

    [Turn over

Marks

2.   **Reference Diagram Q2: A Lowland River Landscape**

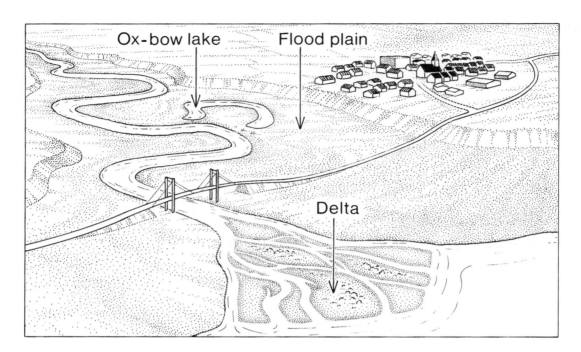

Look at Reference Diagram Q2.

Choose **one** of the named river features shown and **explain** how it was formed.

You may use a diagram(s) to illustrate your answer.

_____

_____

_____

_____

_____

_____

_____

_____

_____

3

3. **Reference Diagram Q3A: Weather Station Symbol for Aberdeen 12 noon, 17th December**
   Reference Diagram Q3B: Advertisement for Football Match

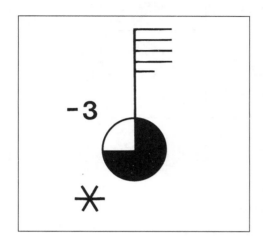

Look at Reference Diagrams Q3A and Q3B.

On Saturday morning the referee decided to postpone this game.

Referring to the weather conditions, give reasons for his decision.

_____

_____

_____

_____

_____

_____

4

[Turn over

Marks

4.   **Reference Diagram Q4A: Causes of Sea Pollution**

A	Sewage and industrial waste	40%
B	Wind-blown gases and particles from industry	35%
C	Oil spills from tankers	15%
D	Dumping at sea	10%

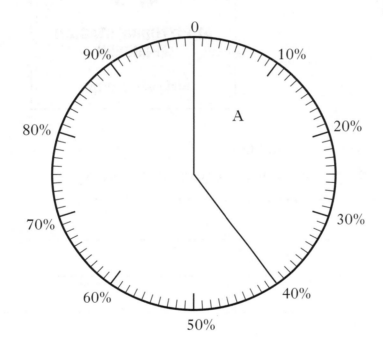

(a)   Complete the pie chart to show the information given in Reference Diagram Q4A.   3

Marks

4. (continued)

Reference Diagram Q4B: Methods of reducing Sea Pollution

A	Enforce laws banning dumping at sea
B	Ensure that sewage is treated before it goes out to sea

(*b*) Which **one** of the above measures do you think would be the best method of reducing sea pollution?

Choice: **A** or **B** _____

Give reasons for your choice.

_____

_____

_____

_____

_____

_____   **3**

[Turn over

5.   **Reference Diagram Q5A: Selected Climate Regions**

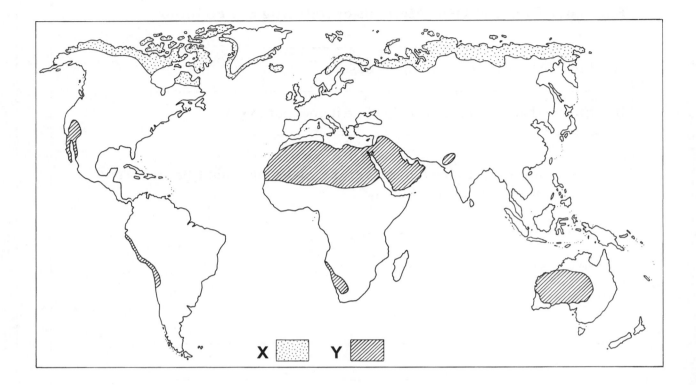

5. (continued)

Reference Diagram Q5B: Climate Graphs for Selected Regions

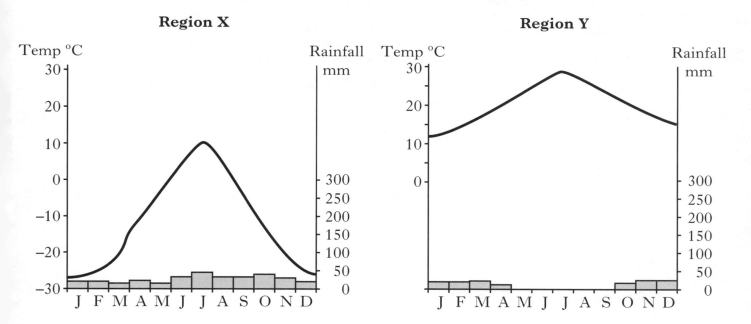

Look at Reference Diagrams Q5A and Q5B.

Both regions X and Y are sparsely populated.

Referring to the climate information, **explain** why it is difficult for people to live and work in each of these regions.

Region X _____

_____

_____

_____

Region Y _____

_____

_____

_____

DO NOT WRITE IN THIS MARGIN

KU	ES

Marks

4

KU | ES

Marks

6.   **Reference Diagram Q6: Selected Farm Data**

Average Temperature	January 5°C, July 16°C
Annual Precipitation	600–800 mm
Relief	Flat, gently sloping land
Soils	Alluvial soils on floodplain
Location	5 km to nearest town

Look at Reference Diagram Q6.

The conditions shown might be suitable for either dairy farming or arable farming.

Which type of farming do you think is more likely?

Tick (✓) your choice.

Dairy farming ☐   Arable farming ☐

Give reasons for your choice.

_____

_____

_____

_____

_____

4

Marks

7.   **Reference Diagram Q7: A Modern Industrial Landscape**

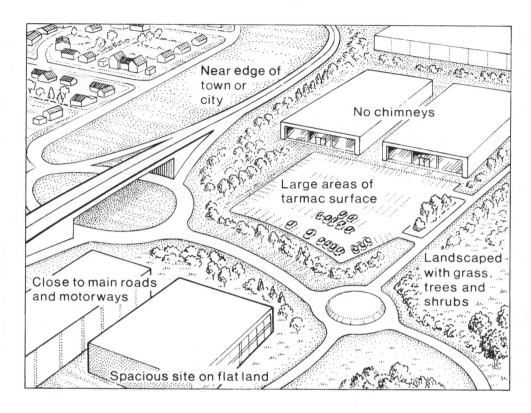

Look at Reference Diagram Q7.

Explain why some of the labelled features are typical of a modern industrial landscape.

_____

_____

_____

_____

_____

_____

4

[Turn over

Marks

8. **Reference Diagram Q8: Reducing Traffic Congestion**

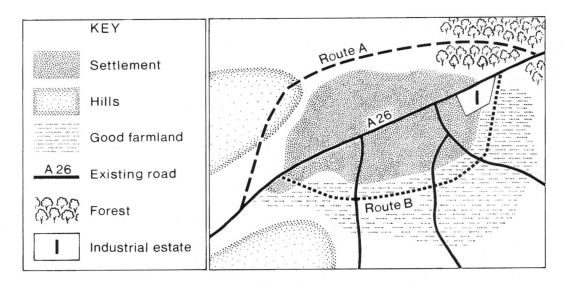

Look at Reference Diagram Q8.

One way of reducing traffic congestion is to build a *bypass*.

Look at Reference Diagram Q8 which shows two possible routes round a settlement.

Which route, A **or** B, do you think would be the better choice?

Tick (✓) your choice.

Route A ☐   Route B ☐

Give reasons for your choice.

_____

_____

_____

_____

_____

_____   **4**

[Turn over for Question 9 on *Page sixteen*

Marks

9.   **Reference Diagram Q9A: Factors which influence
Death Rates in Europe**

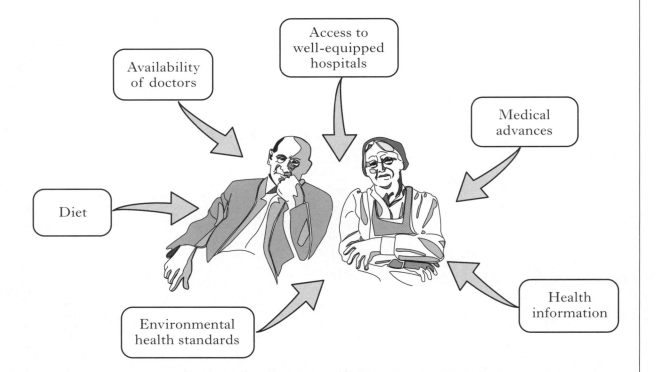

Access to
well-equipped
hospitals

Availability
of doctors

Medical
advances

Diet

Health
information

Environmental
health standards

(*a*)   Study Reference Diagram Q9A.

For any **two** of the factors shown, **explain** how they affect death rates
in Europe.

Factor 1 _____

_____

_____

_____

_____

Factor 2 _____

_____

_____

_____

_____   4

Marks

9. (continued)

Reference Diagram Q9B: Life Expectancy in Italy

Year	Life Expectancy
1950	64
1960	66
1970	70
1980	72
1990	75
2000	78
2005	79

(b)   Look at Reference Diagram Q9B.

Name **two** other techniques which could be used to show the information given.

Give reasons for your choices.

Technique 1   _____

Reason(s)   _____

_____

_____

_____

Technique 2   _____

Reason(s)   _____

_____

_____

_____

4

[Turn over

Marks

10.   **Reference Diagram Q10A: Tied Aid**

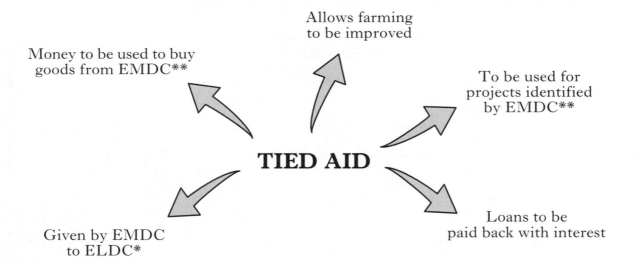

Allows farming
to be improved

Money to be used to buy
goods from EMDC**

To be used for
projects identified
by EMDC**

TIED AID

Loans to be
paid back with interest

Given by EMDC
to ELDC*

*ELDC = Economically Less Developed Country
**EMDC = Economically More Developed Country

(*a*)   Look at Reference Diagram Q10A.

What are the advantages **and** disadvantages of tied aid for ELDCs?

_____

_____

_____

_____

_____

4

Marks

10. (continued)

Reference Diagram Q10B: Education in ELDCs

(*b*)   Look at Reference Diagram Q10B.

"Education is the best way to improve living conditions in ELDCs."

Do you agree fully with this statement?

Give reasons for your answer.

_____

_____

_____

_____

_____

_____   **3**

[Turn over for Question 11 on *Page twenty*

KU | ES

Marks

11.   **Reference Diagram Q11: Location of UK and India**

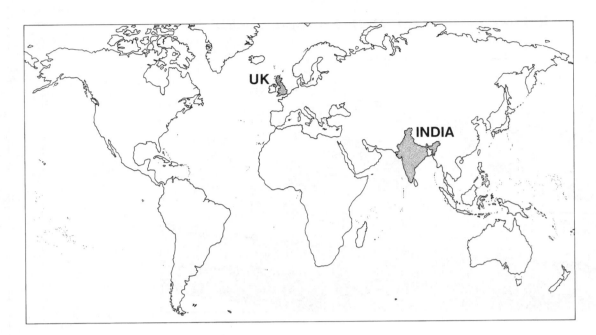

The following statements are about Development and Trade in the UK or India.

Match the letters to the correct country in the table below.

A   Quotas on exports to Germany and France

B   Exports mainly manufactured goods

C   Agriculture employs 66% of population

D   Energy consumption per capita is low

E   Agriculture employs 2% of population

F   High GNP per capita

UK	India

4

[END OF QUESTION PAPER]

KU | ES

2008

[BLANK PAGE]

C

1260/405

NATIONAL QUALIFICATIONS 2008	FRIDAY, 9 MAY 1.00 PM – 3.00 PM	GEOGRAPHY STANDARD GRADE Credit Level

All questions should be attempted.

Candidates should read the questions carefully. Answers should be clearly expressed and relevant.

Credit will always be given for appropriate sketch-maps and diagrams.

Write legibly and neatly, and leave a space of about one centimetre between the lines.

All maps and diagrams in this paper have been printed in black only: no other colours have been used.

PB 1260/405 6/18070

Extract No 1656/104

1:50 000 Scale
Landranger Series

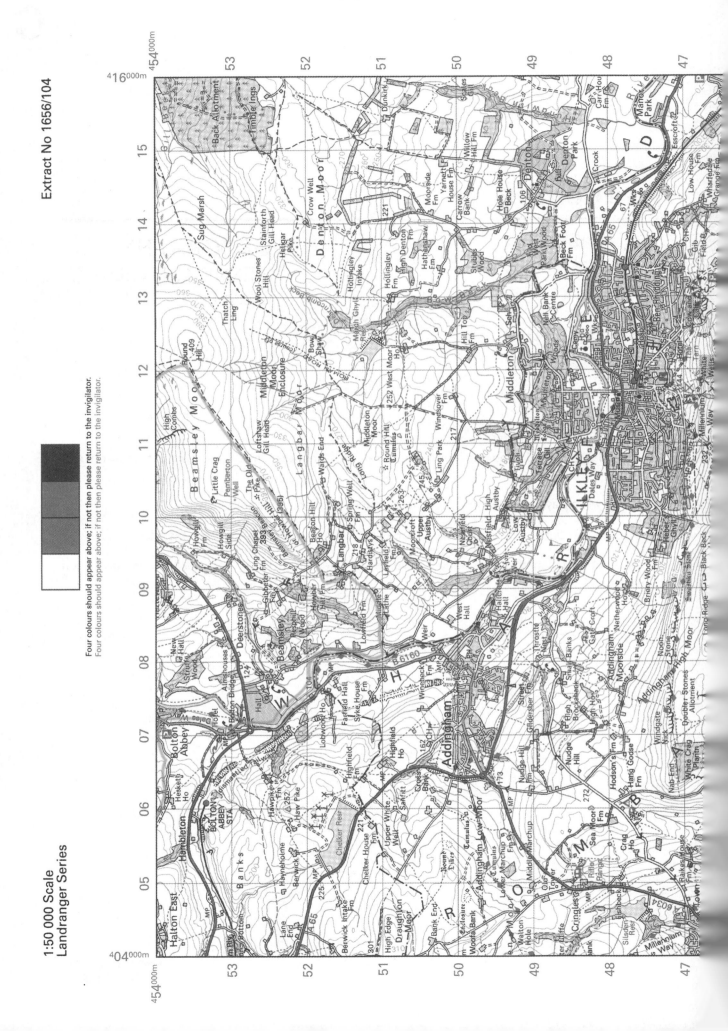

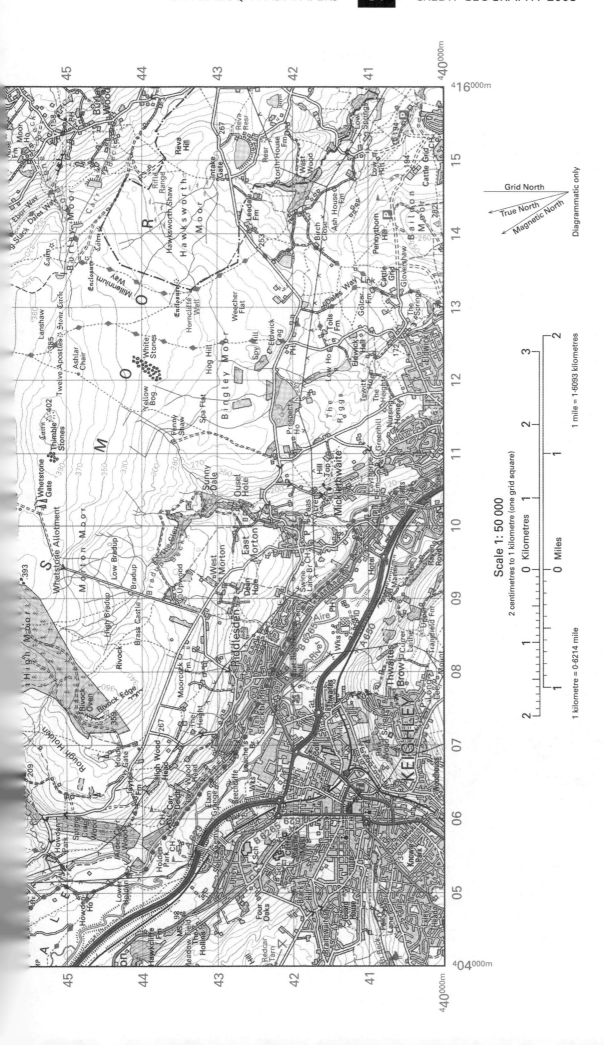

Scale 1:50 000

2 centimetres to 1 kilometre (one grid square)

Grid North

True North

Magnetic North

Diagrammatic only

Extract produced by Ordnance Survey 2007.
© Crown copyright 2006. All rights reserved.

1.   **Reference Diagram Q1**

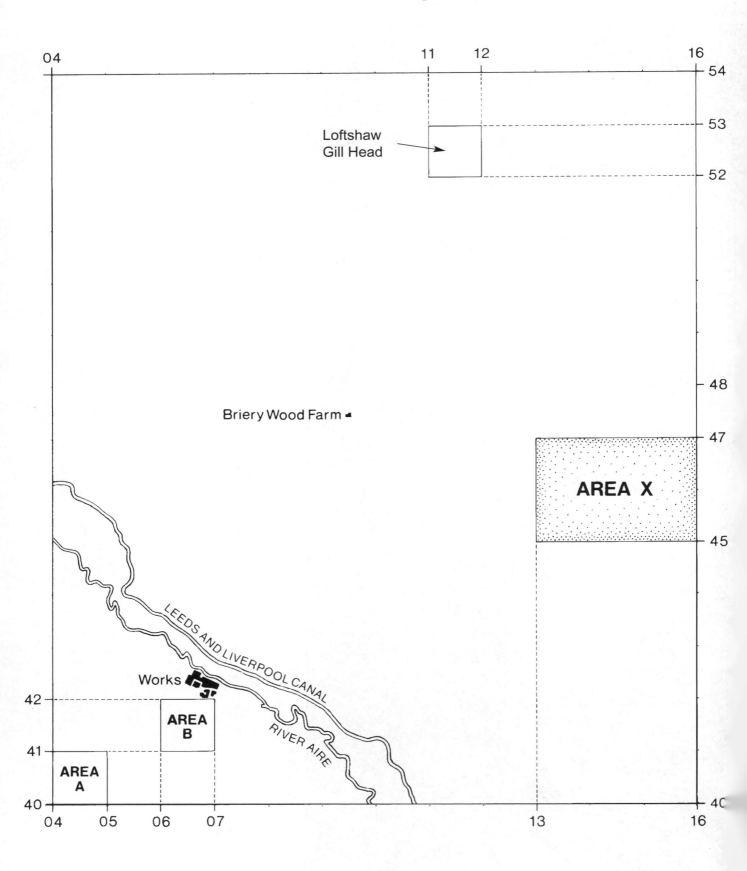

Marks

KU	ES

1. (continued)

This question refers to the OS Map Extract (No 1656/104) of the Ilkley area and Reference Diagram Q1.

(*a*) Describe the **physical** features of the River Aire **and** its valley from 041450 to 099400.

Your answer should **not** refer to the Leeds and Liverpool canal.   **4**

(*b*) The valley which runs south east from Loftshaw Gill Head in square 1152 is a "V" shaped river valley. Explain how this is likely to have been formed.

You may use diagrams to illustrate your answer.   **4**

(*c*) Find Briery Wood Farm at 095474.

What are the advantages **and/or** disadvantages of this location for a farm?   **4**

(*d*) Why might there be conflicts between the various land uses in Area X?   **6**

(*e*) Describe the differences between the urban environments of Area A (0440) and Area B (0641).   **5**

(*f*) Find the works in grid square 0642.

Why is this a good site for the works?   **5**

[Turn over

Mark

KU

2.   **Reference Diagram Q2: A Lowland Landscape in Scotland**

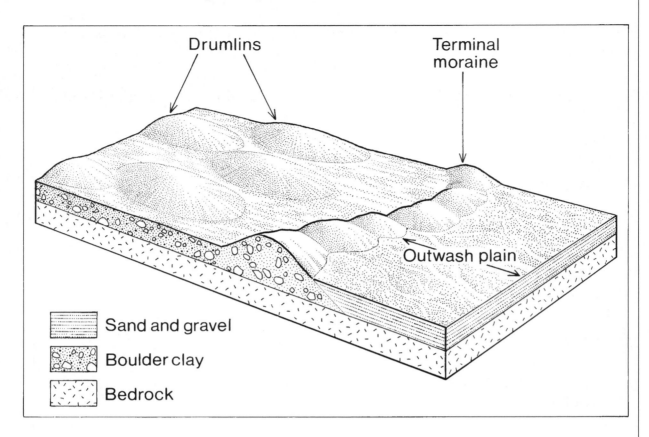

Look at Reference Diagram Q2.

Choose any **two** of the features shown in the diagram and **explain** how they were formed.

You may use diagram(s) to illustrate your answer.

5

3.   **Reference Diagram Q3: European Synoptic Chart for noon, 8th July**

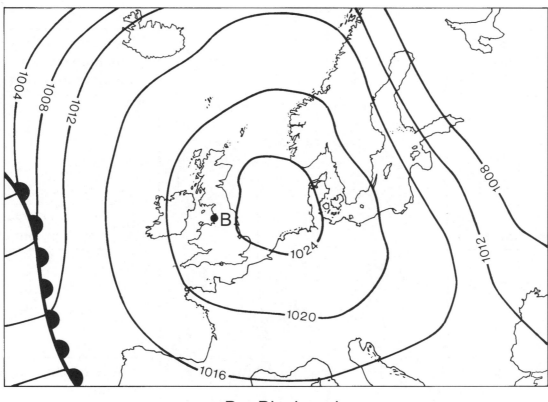

● B – Blackpool

Look at Reference Diagram Q3.

On July 8th Mr McCormack is taking his young family to Blackpool for a one week holiday.

Do you think the weather conditions will be favourable for them?

Give reasons your answer.

5

[Turn over

4.   **Reference Diagram Q4: Brazilian Rainforest Facts**

Mineral Resources

Native American Indians living on reserves

Largest number of plant and animal species of any natural region

Valuable hardwood timber

Hydro-electric power potential

In-migration of settlers

Cattle ranching and plantation agriculture

Medicines derived from plants

Trans-Amazonian highway opening up remote areas

Trees convert carbon dioxide to oxygen

Ma

KU

"Cutting down the rainforest in Brazil will affect the whole world more than it will affect Brazil itself."

(Statement by an environment spokesperson)

Look at Reference Diagram Q4 and the statement above.

To what extent do you agree with the statement?

Give reasons for your answer.

5.   **Reference Diagram Q5A: Location of Housing Types**

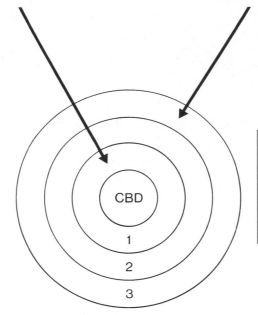

Key to Zones

CBD Central Business District
  1   19th Century Housing
  2   1930s Inter War Housing
  3   Late 20th Century Housing

	Marks	
	KU	ES

(a) Look at Reference Diagram Q5A.

Explain in detail why there are different types of houses in Zones 1 and 3. | 6 |

Reference Diagram Q5B: Statement

"There have been many changes in the CBD."

(b) Look at Reference Diagram Q5B.

What techniques could a group of Geography students use to gather information on changes in the Central Business District (CBD)?

Give reasons for your chosen techniques.   **5**

Marks

KU | E

6.   **Reference Diagram Q6A: Eurocentral**

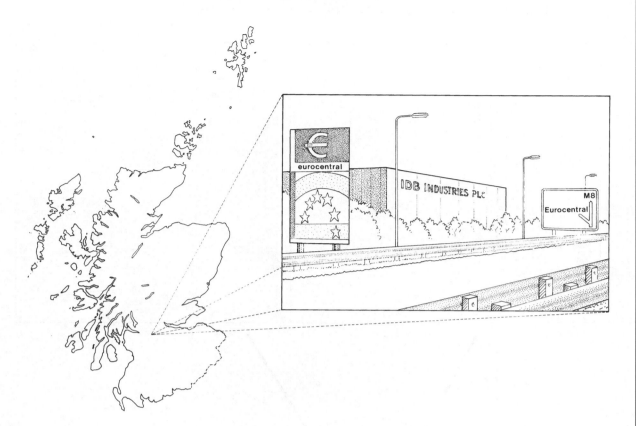

Reference Diagram Q6B: Extract from a News Report

"Eurocentral is a 260 hectare greenfield site where new large-scale industrial development is planned.

This is expected to have a significant impact on the surrounding communities, where old industry has been in decline."

(*a*) Look at Reference Diagrams Q6A and Q6B.

Describe the benefits **and** problems which new, large-scale industrial development may bring to areas such as this.

6

Marks

KU	ES

6. (continued)

Reference Diagram Q6C: North Lanarkshire—Selected Industrial Statistics

Table 1: Employment Categories

Public and other services	30 700
Retailing and wholesale	24 800
Manufacturing	18 700
Finance and business	14 500

Table 2: Unemployment 1996–2002

1996	12 500
1997	10 500
1998	10 000
1999	8500
2000	8000
2001	7500
2002	7000

(b) Look at Reference Diagram Q6C.

What other techniques could be used to present the data shown above?

Give reasons for your choices.

5

[Turn over

Marks

KU E

7. **Reference Diagram Q7A: Demographic Transition Model**

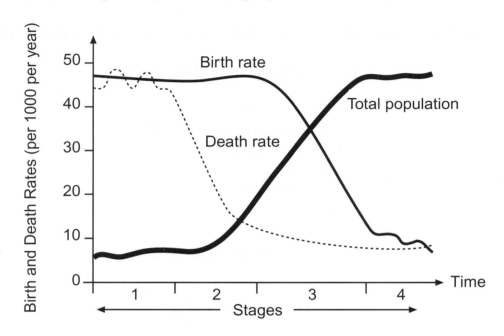

(a) Look at Reference Diagram Q7A.

Describe in detail the changes shown on the Demographic Transition Model from Stage 1 to Stage 4.

Reference Diagram Q7B: Selected Population Data

Country	Crude Birth Rate per 1000	Crude Death Rate per 1000	Natural Increase per 1000
India	23	8	15
Nigeria	38	14	24
UK	10·8	10·1	0·7
USA	14·1	8·3	5·8

(b) Look at Reference Diagram Q7B.

Choose **one** country shown on the table.

Suggest reasons for its **rate of natural increase**.

4

Marks
KU | ES

8.   **Reference Diagram Q8A: Japan's Exports**

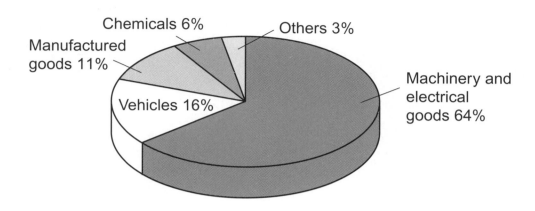

Reference Diagram Q8B: Japan's Imports

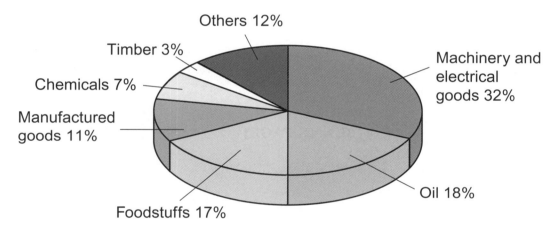

Look at Reference Diagrams Q8A and Q8B.

(*a*) What are the main differences between Japan's exports and imports?   3

(*b*) Japan has a large trade surplus and is the world's second biggest trading nation.

  Explain why Japan depends on world trade for the success of its economy.   3

[END OF QUESTION PAPER]

[BLANK PAGE]

[BLANK PAGE]

FOR OFFICIAL USE

KU   ES

Total Marks

1260/403

NATIONAL
QUALIFICATIONS
2009

WEDNESDAY, 27 MAY
10.25 AM–11.50 AM

GEOGRAPHY
STANDARD GRADE
General Level

G

Fill in these boxes and read what is printed below.

Full name of centre

Town

Forename(s)

Surname

Date of birth
Day Month Year

Scottish candidate number

Number of seat

1 Read the whole of each question carefully before you answer it.

2 Write in the spaces provided.

3 Where boxes like this ☐ are provided, put a tick ✓ in the box beside the answer you think is correct.

4 Try all the questions.

5 Do not give up the first time you get stuck: you may be able to answer later questions.

6 Extra paper may be obtained from the invigilator, if required.

7 Before leaving the examination room you must give this book to the invigilator. If you do not, you may lose all the marks for this paper.

Extract No 1742/140

1:50 000 Scale
Landranger Series

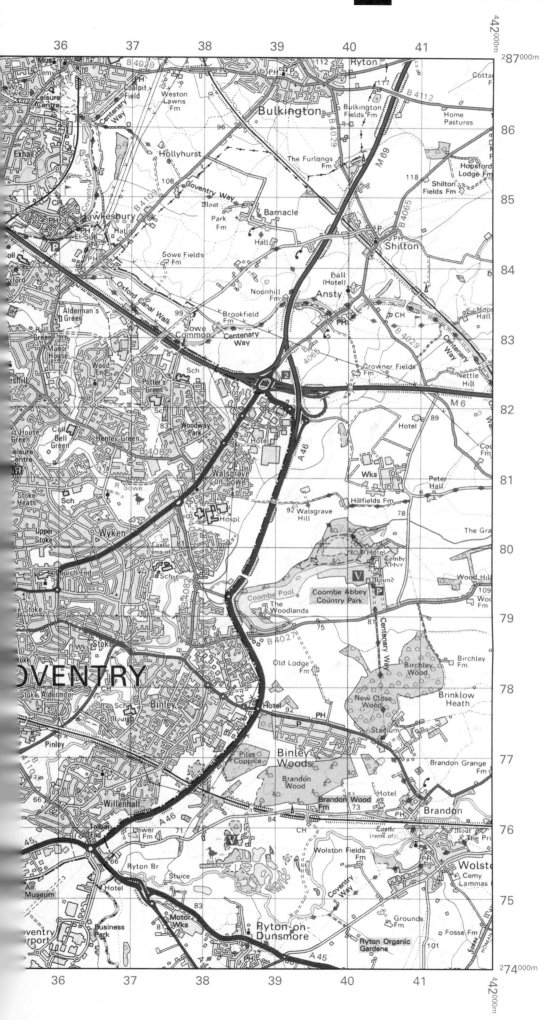

Scale 1 : 50 000

2 centimetres to 1 kilometre (one grid square)

1 mile = 1·6093 kilometres

1 kilometre = 0·6214 mile

Ordnance Survey, OS, the OS Symbol and Landranger are registered trademarks of Ordnance Survey, the national mapping agency of Great Britain. Reproduction in whole or in part by any means is prohibited without the prior written permission of Ordnance Survey. **For educational use only.**

Grid North

True North

Magnetic North

Diagrammatic only

1.

Reference Diagram Q1A

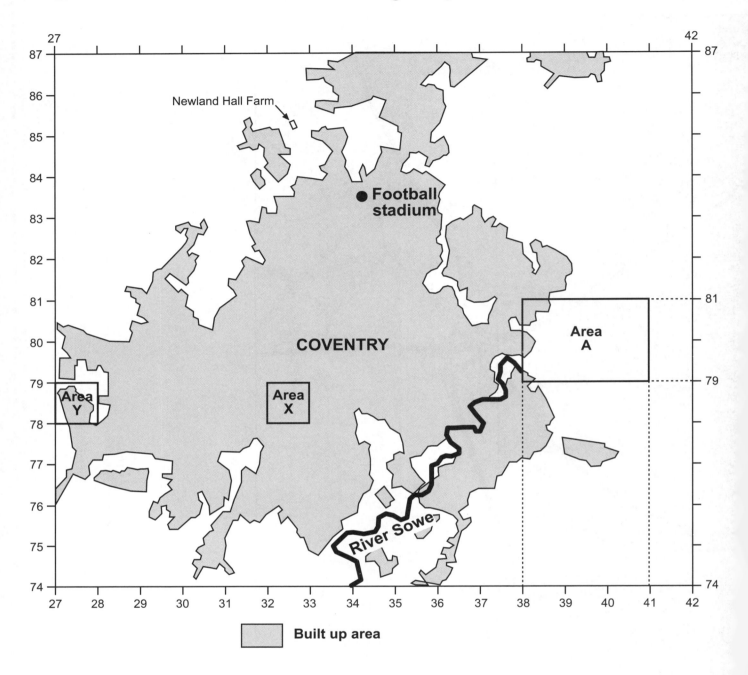

Built up area

Marks

1. (continued)

Look at the Ordnance Survey Map Extract (No 1742/140) of Coventry and Reference Diagram Q1A on *Page two*.

(*a*) Describe the **physical** features of the River Sowe **and** its valley from 378796 to 340740. (This section of the river is shown on Reference Diagram Q1A.)

_____

_____

_____

_____

_____

_____

4

(*b*) Find Area A on Reference Diagram Q1A and the map extract.

What are the advantages **and** disadvantages of this area for a Country Park?

_____

_____

_____

_____

_____

_____

4

[Turn over

Marks

1. (continued)

(*c*)   Give map evidence to show that part of Coventry's Central Business District (CBD) is located in grid square 3379.

_____

_____

_____

_____

_____

_____   **4**

(*d*)   A young couple want to buy a house in Coventry. They have decided to buy a house in either Area X (grid square 3278) or in Area Y (grid square 2778).

Using map evidence, which area, X or Y, would you advise them to choose?

Give reasons for your answer.

Choice: _____

Reasons _____

_____

_____

_____

_____   **4**

DO NOT WRITE IN THIS MARGIN

KU ES

Marks

1. (continued)

(*e*) Newland Hall Farm is located in map square 3285.

What are the advantages **and** disadvantages of this location for farming?

Advantages ______________________________

____

____

____

Disadvantages ______________________________

____

____

______________________________ **4**

[Turn over

1. **(continued)**

Reference Diagram Q1B: Land Use at Grid Reference 343835 in 1999

Reference Diagram Q1C: Land Use at Grid Reference 343835 in 2009

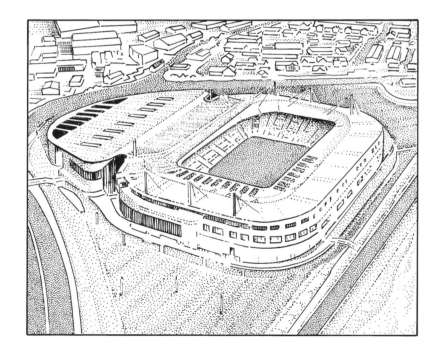

KU | ES

Marks

1. (continued)

(f)   Look at Reference Diagrams Q1B and Q1C on *Page six*.

A new football ground was built for Coventry City Football Club on the site of a former gas works at 343835.

Do you think this change in land use will have brought benefits to the area?

Using map evidence, give reasons for your answer.

_____

_____

_____

_____

_____

_____   **3**

[Turn over

Marks

2. **Explain** how **either** an oxbow lake **or** a waterfall is formed.

You may use diagrams to illustrate your answer.

_____

_____

_____

_____

_____

_____

_____

_____

_____

4

3. **Reference Diagram Q3A: Synoptic Chart for Bordeaux
at noon on 30 March 2007**

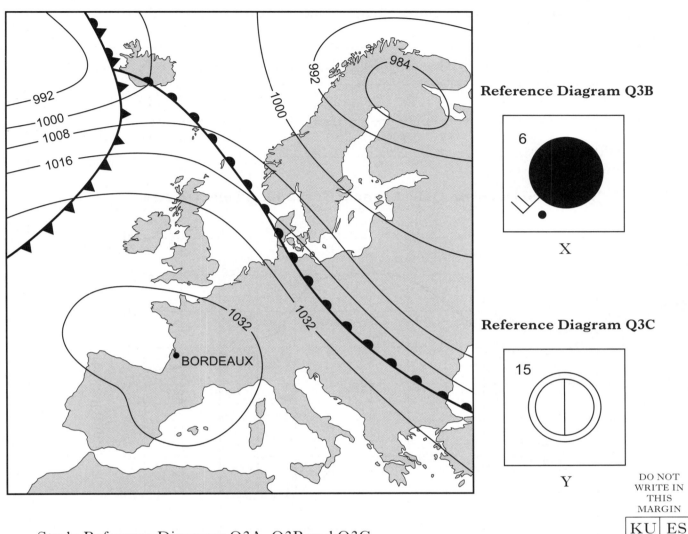

Reference Diagram Q3B

X

Reference Diagram Q3C

Y

Study Reference Diagrams Q3A, Q3B and Q3C.

Which weather station circle, X or Y, is the correct one for Bordeaux at noon on 30 March 2007?

Tick (✓) your choice.

Station circle X   ☐   Station circle Y   ☐

Give reasons for your choice.

_____

_____

_____

_____

_____

_____

	KU	ES
Marks		

4

	KU	ES

Marks

4.   Reference Diagram Q4A: Climate Statistics for a Selected Area

	J	F	M	A	M	J	J	A	S	O	N	D
Temperature (°C)	12	13	15	19	23	25	25	24	17	16	13	12
Rainfall (mm)	66	58	57	57	40	30	20	26	46	68	80	70

Reference Diagram Q4B: Climate Graph for a Selected Area

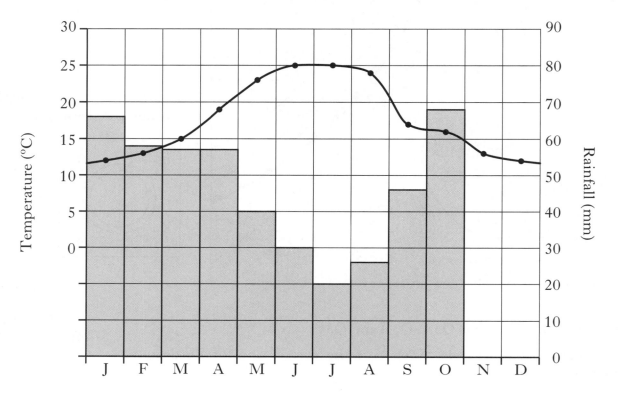

(a)   Look at the table (Reference Diagram Q4A) at the top of the page. Use this data to complete the climate graph.

2

(b)   Describe, **in detail**, the climate shown in Reference Diagram Q4B.

_____

_____

_____

_____

_____

4

Marks

5.   **Reference Diagram Q5: Desertification**

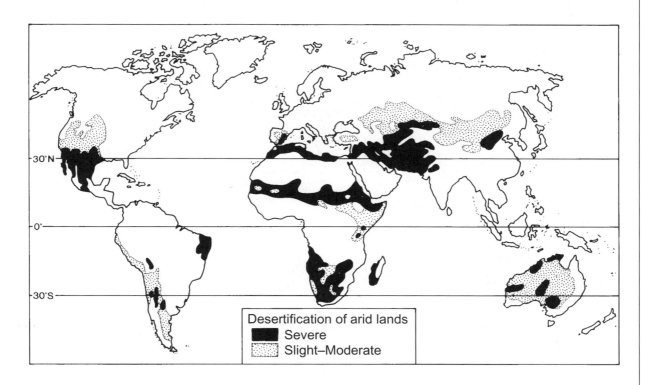

Describe the main causes of desertification.

_____

_____

_____

_____

_____

_____

3

[Turn over

6. **Reference Diagram Q6A: Location of Prologis Business Park, Coventry**

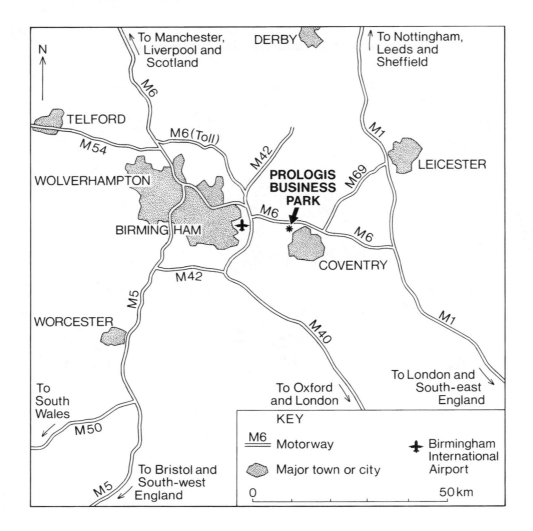

Reference Diagram Q6B: Business Park

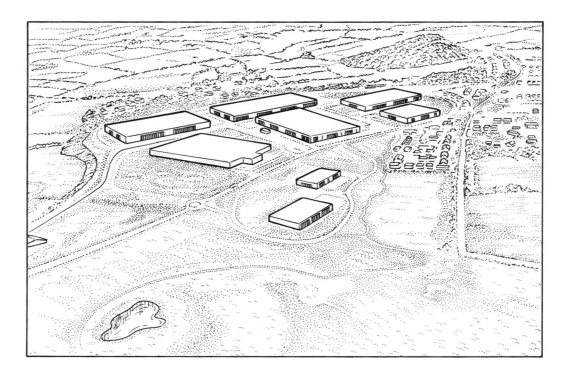

Marks

6. (continued)

(*a*) Study Reference Diagrams Q6A and Q6B.

Why is this a good location for a business park?

_____

_____

_____

_____

_____

_____

_____

4

(*b*) Identify **two** gathering techniques which a group of pupils could use to find out how the Business Park affects the area around Coventry.

Give reasons for your answer.

Technique 1 _____

_____

Reason(s) _____

_____

_____

Technique 2 _____

_____

Reason(s) _____

_____

_____

4

[Turn over

Marks

**7.   Reference Diagram Q7: Solutions to Traffic Congestion in
a British City Centre**

Congested Traffic

One-way street
system

Ring road

Flexi-time
working _____ SOLUTIONS TO
CONGESTION _____ Parking restrictions
(eg meters, wardens,
double yellow lines)

Multi-storey
car parks

Park and ride
schemes

Look at Reference Diagram Q7.

Which **two** solutions would be most effective in reducing traffic congestion
in a British city centre?

Give reasons for your choices.

Solution 1 _____

Reasons _____

_____

_____

Solution 2 _____

Reasons _____

_____

_____

4

Marks

8. **Reference Diagram Q8: Features of the European Union (EU)**

trade barriers with
non members

free trade between
members

currency
(Euros)

economic help to
selected regions

free movement of workers
between member countries

Look at Reference Diagram Q8.

"Membership of the European Union (EU) has both advantages and disadvantages **for the UK**."

What are these advantages and disadvantages?

Advantages _____

_____

_____

_____

Disadvantages _____

_____

_____

_____   **4**

[Turn over

9. **Reference Diagram Q9A: Population Structure for Hungary in 2009**

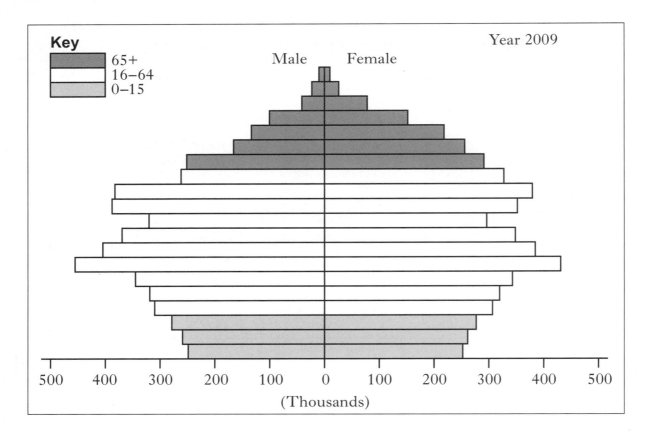

Reference Diagram Q9B: Expected Population Structure for Hungary in 2049

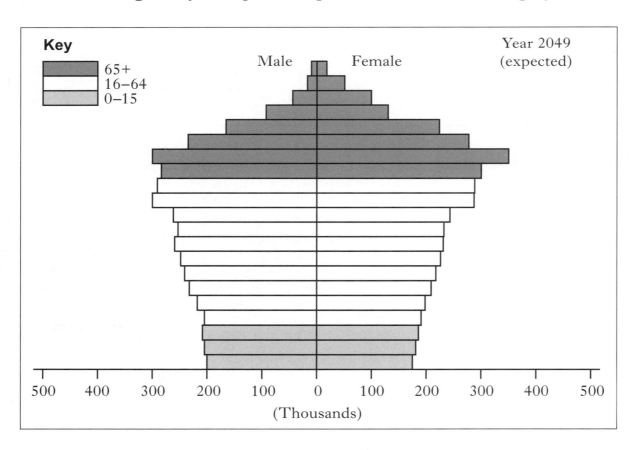

Marks

9. (continued)

Look at Reference Diagrams Q9A and Q9B.

(*a*) Describe the changes which are expected to take place in the population structure of Hungary by 2049.

_____

_____

_____

_____

_____   **3**

Look at Reference Diagram Q9B.

(*b*) What problems might the expected population structure in 2049 create for Hungary?

_____

_____

_____

_____

_____

_____   **4**

[Turn over

Marks

10. **Reference Diagram Q10: United States—Imports and Exports (2000–2005 Average)**

% Imports

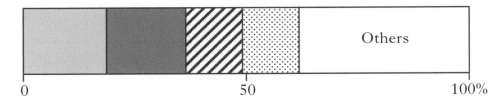

% Exports

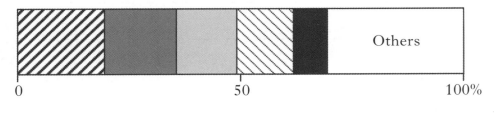

- ☐ Vehicles
- ☐ Fuels
- ☐ Electric & electronic
- ☐ Aircraft
- ☐ Nuclear equipment
- ☐ Scientific equipment

Look at Reference Diagram Q10.

Give **other** processing techniques which could be used to show the information in the diagrams above.

Why are these techniques suitable?

_____

_____

_____

_____

_____

_____

4

KU | ES

Marks

11. **Reference Diagram Q11: Self-help Schemes**

Improved Plough

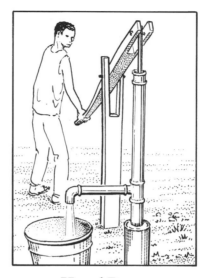

Hand Pump

Stone Lines

Look at Reference Diagram Q11.

Why are self-help schemes suitable for Economically Less Developed Countries (ELDCs)?

_____

_____

_____

_____

_____

3

[END OF QUESTION PAPER]

KU | ES

[BLANK PAGE]

STANDARD GRADE | CREDIT

2009

[BLANK PAGE]

C

1260/405

NATIONAL
QUALIFICATIONS
2009

WEDNESDAY, 27 MAY
1.00 PM – 3.00 PM

GEOGRAPHY
STANDARD GRADE
Credit Level

All questions should be attempted.

Candidates should read the questions carefully. Answers should be clearly expressed and relevant.

Credit will always be given for appropriate sketch-maps and diagrams.

Write legibly and neatly, and leave a space of about one centimetre between the lines.

All maps and diagrams in this paper have been printed in black only: no other colours have been used.

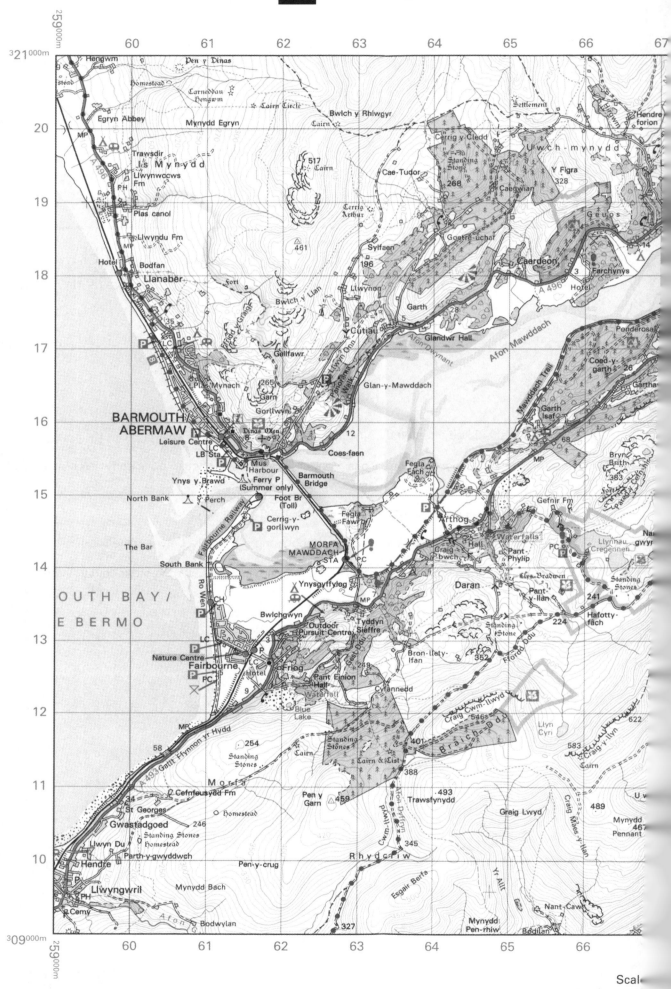

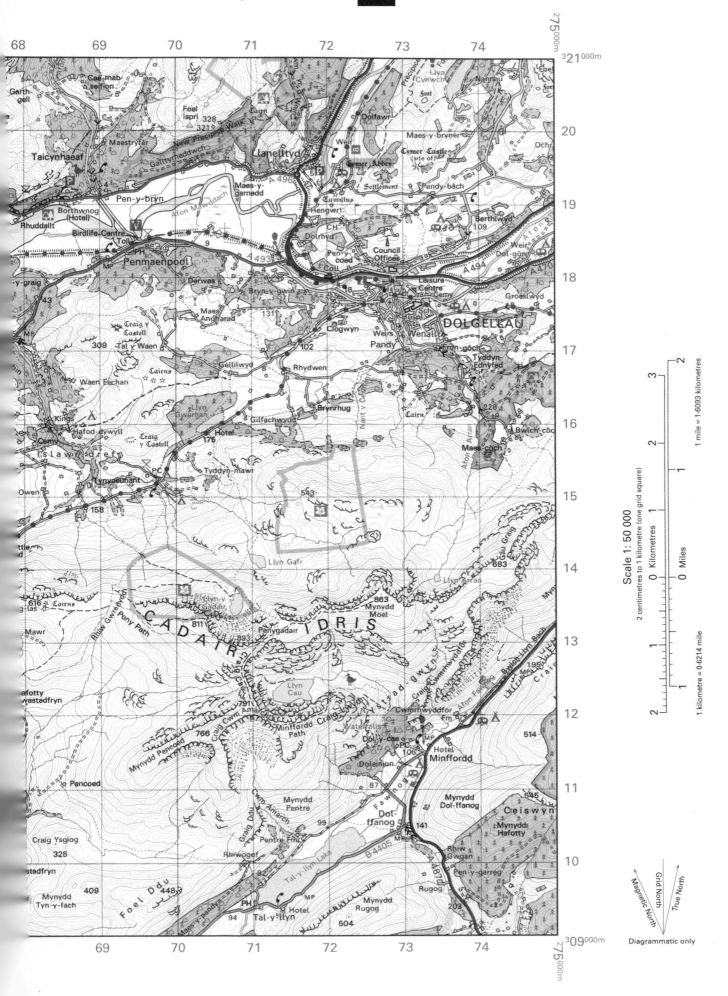

Scale 1: 50 000

2 centimetres to 1 kilometre (one grid square)

1 mile = 1·6093 kilometres

1 kilometre = 0·6214 mile

1.   **Reference Diagram Q1A**

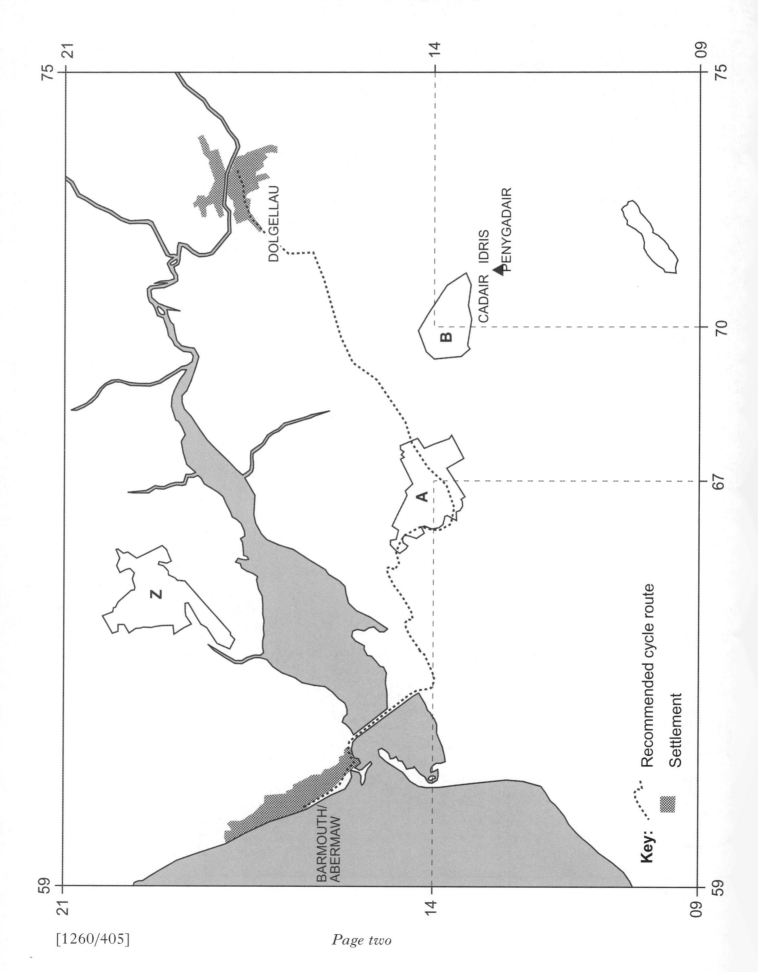

Key: ⋯⋯ Recommended cycle route
        ▣ Settlement

Marks

KU	ES

1. (continued)

This question refers to the OS Map Extract (No 1743/124) of the Dolgellau area and Reference Diagram Q1A on *Page two*.

(*a*) (i) Match each of the features named below with the correct grid reference.

     Features: **pyramidal peak**; **corrie**; **truncated spur**; **hanging valley**.

     Choose from grid references: 711130, 723125, 715123, 733110.

  (ii) **Explain** how **one** of the features listed in (*a*)(i) was formed.

     You may use diagrams to illustrate your answer.

3

4

(*b*)   **Reference Diagram Q1B: Selected Settlement Functions**

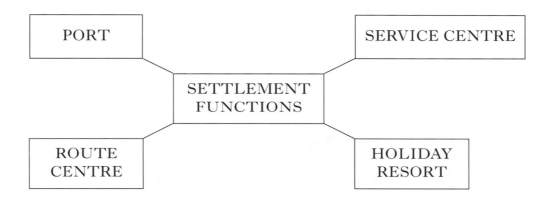

Look at Reference Diagram Q1B.

Using map evidence, compare the main functions of Barmouth and Dolgellau. You must refer to both similarities **and** differences in your answer.

6

(*c*) Find area Z shown on Reference diagram Q1A.

Why is this area suited to commercial forestry?

4

[Turn over

1. (continued)

(*d*) A family cycling from Dolgellau to Barmouth have been recommended to follow the route shown on Reference Diagram Q1A.

What are the advantages and disadvantages of taking this route?

(*e*) A group of students wish to carry out a study to compare the two National Trust areas A and B, shown on Reference Diagram Q1A.

Describe gathering techniques which could be used to obtain relevant information.

Give reasons for your choice of techniques.

2.   **Reference Diagram Q2: A V-shaped Valley**

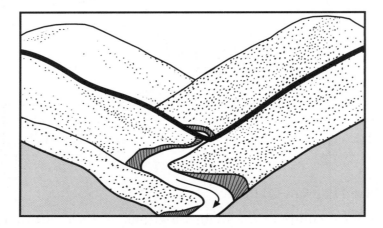

Study Reference Diagram Q2 above.

Explain the formation of a V-shaped valley.

You may use diagrams to illustrate your answer.

4

3.   **Reference Diagram Q3: Synoptic Chart for 15th February 2007**

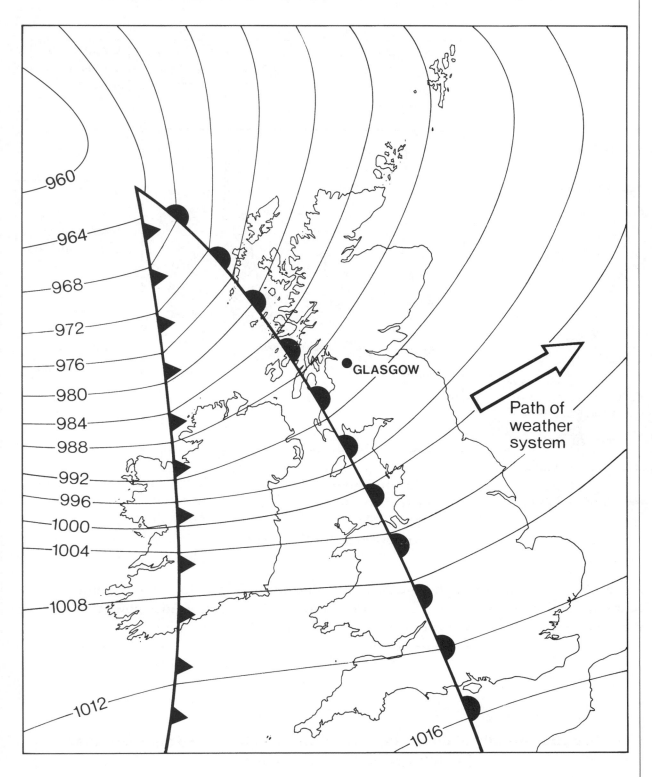

Study Reference Diagram Q3.

Explain the changes which will take place in the weather in Glasgow over the next 24 hours.

6

[Turn over

4.   **Reference Diagram Q4**

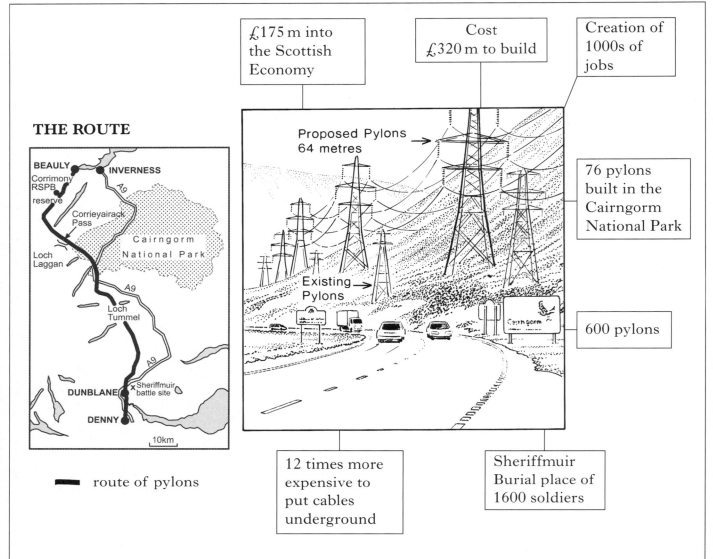

It is proposed to build a high voltage power line from Beauly in the Highlands to Denny in Stirlingshire, transferring energy from wind, wave and tidal power. This green energy would replace power from coal and gas fired power stations.

Study Reference Diagram Q4.

"The benefits brought about by the power line will be more important than the damage to the countryside."

Do you agree fully with this statement?

Give reasons for your answer.

Ma

KU

5.　**Reference Diagram Q5:　Land Use in a Scottish City**

Suburbs　　　　　　　**Central Business District**

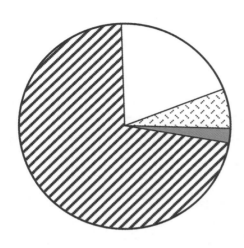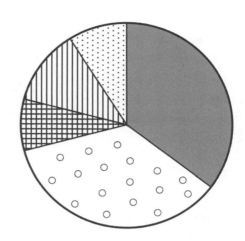

| Residential | Hotels | Public Buildings | Shops |
| Open space | Offices | Entertainment | Industry |

Look at Reference Diagram Q5.

Give reasons for the patterns of the land use shown in both the suburbs and the Central Business District.

6

[Turn over

6.   **Reference Diagram Q6A: Farming Landscape in 1950**

Few big machines

Low yields per hectare

Many farm workers

Small farms

Small irregular fields

Reference Diagram Q6B: Farming Landscape in 2005

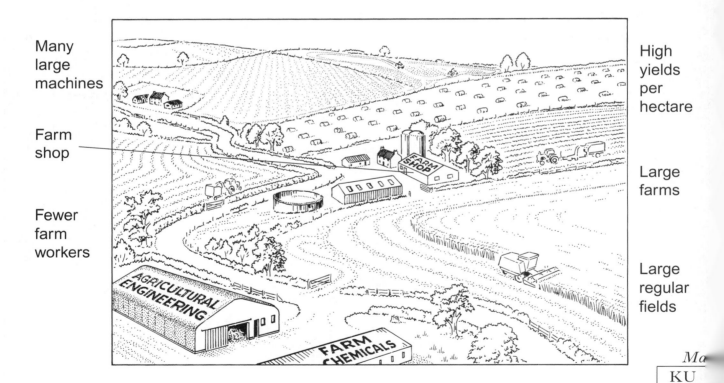

Many large machines

High yields per hectare

Farm shop

Fewer farm workers

Large farms

Large regular fields

Look at Reference Diagrams Q6A and Q6B.

"The changes in farming since 1950 have brought more benefits than problems."

Do you agree fully with this statement?

Give reasons for your answer.

7.

Reference Diagram Q7A: Factors influencing the Location of a Science Park

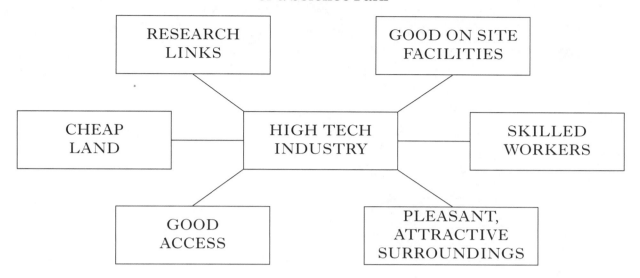

Reference Map Q7B: Location of University of Southampton Science Park

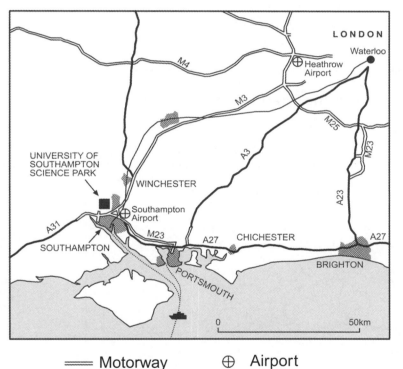

—— Motorway   ⊕ Airport

—— A class road

—— Railway   ⌇ Ferry route

● Railway station   ▓ Built up area

Reference Diagram Q7C: Site of University of Southampton Science Park

Look at Reference Diagrams Q7A, Q7B and Q7C.

In your opinion, which **three** factors were most important in the location of the University of Southampton Science Park?

Give reasons to support your choice.

	Marks	
	KU	ES
		5

Mark

KU

8.   **Reference Diagram Q8A: Population Distribution Map of Europe**

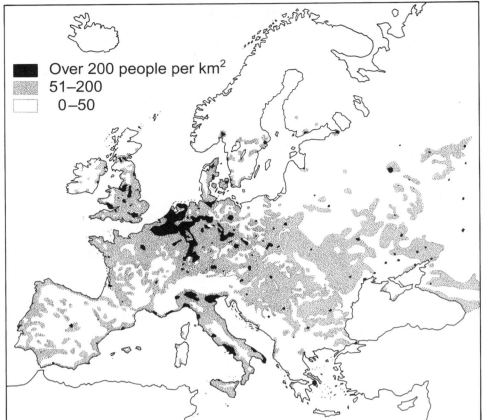

Reference Diagram Q8B: Relief and International Boundaries of Europe

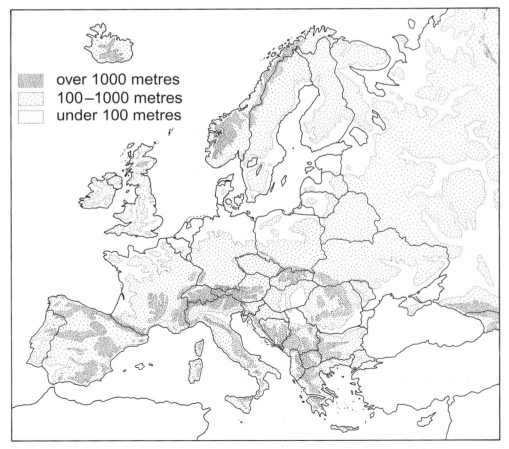

Study Reference Diagrams Q8A and Q8B.

Describe in detail the population distribution in Europe.

Marks

KU	ES

9.   **Reference Diagram Q9A: Change in worldwide rural and urban populations**

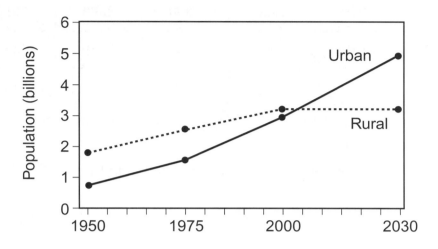

Reference Diagram Q9B: Population living in urban areas

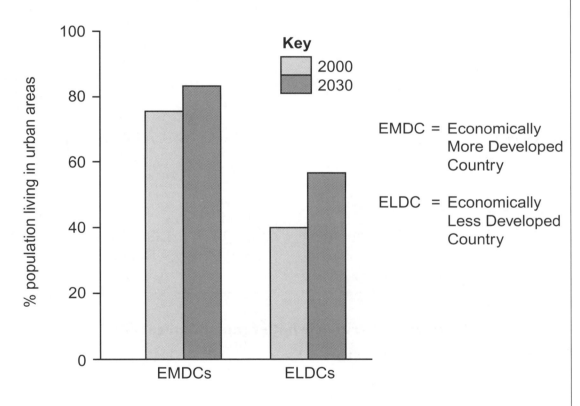

Look at Reference Diagrams Q9A and Q9B.

(a) Describe the trends shown in both graphs.

(b) **Explain** why urban populations are expected to continue increasing.

5   4

[Turn over

Marks

KU

10.   **Reference Diagram Q10: Percentages of Population aged over 65**

Country	1960	1970	1980	1990	2000	2010
Japan	6·0	7·0	9·0	12·0	16·0	22·0
Nigeria	2·3	2·4	2·5	2·5	2·7	3·2

Study Reference Diagram Q10.

What other techniques could be used to show the information in the table?

Give reasons for your choices.

11.   **Reference Diagram Q11: Different Types of Aid in ELDCs**

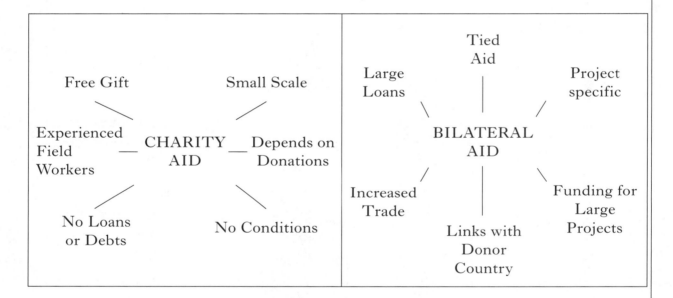

"Aid from charities is better than bilateral aid."

Do you agree fully with this statement?

Give reasons for your answer.

[END OF QUESTION PAPER]

[BLANK PAGE]

FOR OFFICIAL USE

G

KU   ES

Total Marks

1260/403

NATIONAL
QUALIFICATIONS
2010

THURSDAY, 6 MAY
10.25 AM–11.50 AM

GEOGRAPHY
STANDARD GRADE
General Level

Fill in these boxes and read what is printed below.

Full name of centre

Town

Forename(s)

Surname

Date of birth

| Day | Month | Year | Scottish candidate number | Number of seat |

1   Read the whole of each question carefully before you answer it.

2   Write in the spaces provided.

3   Where boxes like this ☐ are provided, put a tick ✓ in the box beside the answer you think is correct.

4   Try all the questions.

5   Do not give up the first time you get stuck: you may be able to answer later questions.

6   Extra paper may be obtained from the Invigilator, if required.

7   Before leaving the examination room you must give this book to the Invigilator. If you do not, you may lose all the marks for this paper.

Extract No 1785/90

1:50 000 Scale
Landranger Series

Four colours should appear above; if not then please return to the invigilator.
Four colours should appear above; if not then please return to the invigilator.

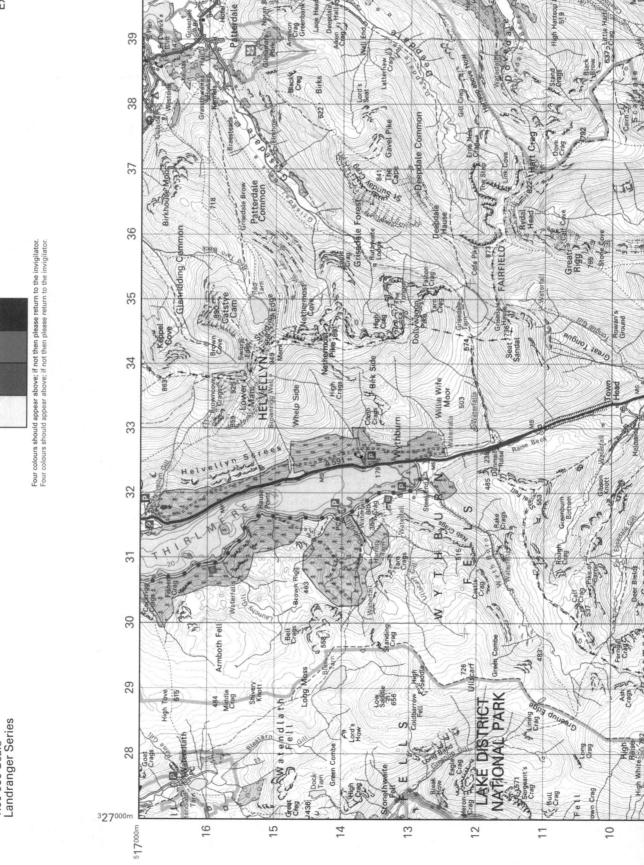

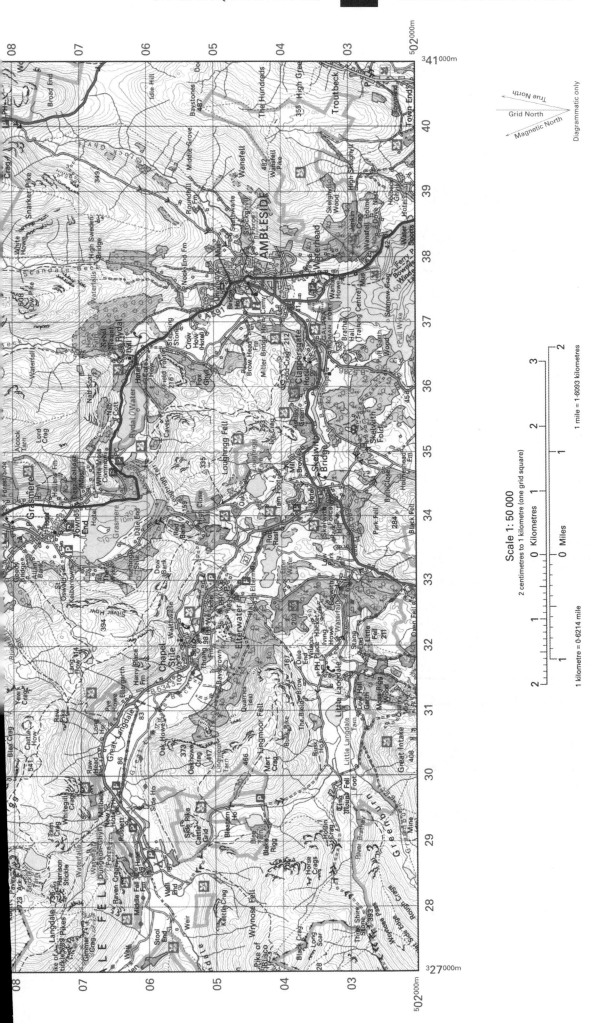

Scale 1 : 50 000

2 centimetres to 1 kilometre (one grid square)

1 kilometre = 0·6214 mile

1 mile = 1·6093 kilometres

Extract produced by Ordnance Survey 2009. Licence: 100035658

1.   **Diagram Q1A: The Ambleside Area**

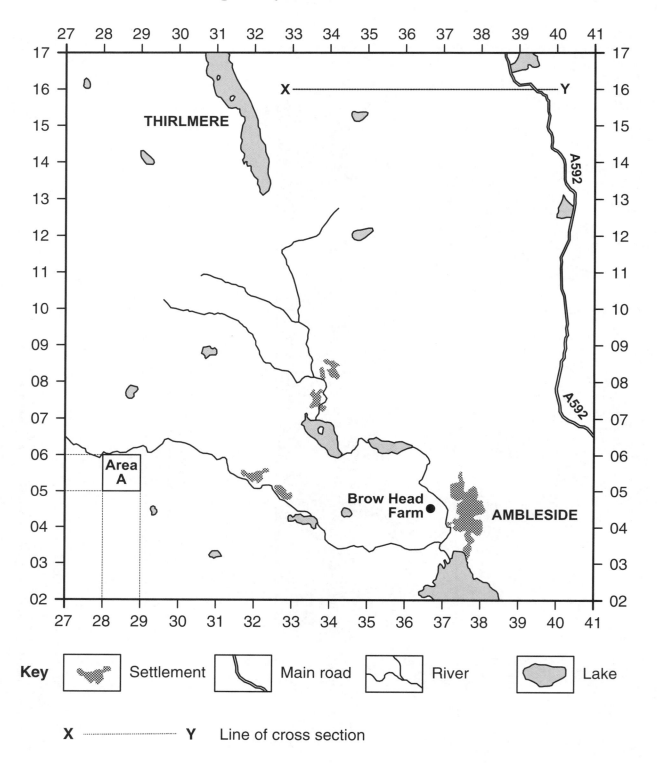

1. (continued)

Look at the Ordnance Survey Map Extract (No 1785/90) of the Ambleside area and Diagram Q1A on *Page two*.

Diagram Q1B: Cross Section from X (330160) to Y (400160)

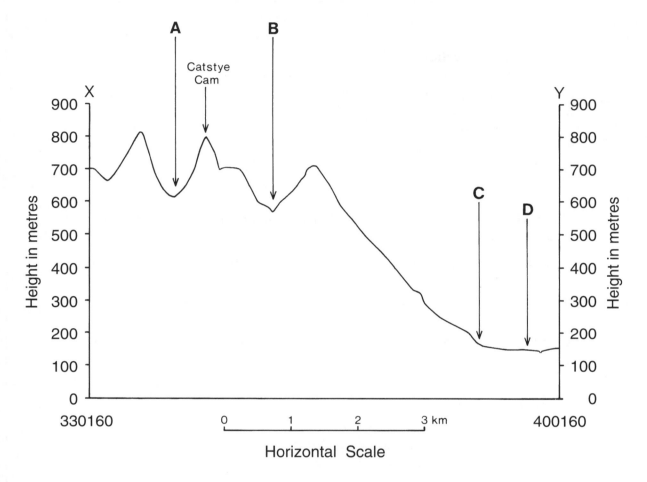

(*a*) Look at Diagram Q1B and Diagram Q1A. Find cross section XY on the map extract.

Match the features A, B, C and D on the cross section XY with the correct description in the table below.

Feature	Letter
A592	
Brown Cove	
Mixed wood	
Red Tarn Beck	

Marks

3

KU | ES

Marks

1. (continued)

(b) (i) Match the glacial features in the table with the grid references below.

2807   3115   3108   3006

Glacial Feature	Grid Reference
Hanging valley	
Corrie with tarn	
Misfit stream	
Ribbon lake	

3

(ii) **Explain** how **one** of the glacial features named in the table above was formed.

You may use a diagram(s) to illustrate your answer.

_____

_____

_____

_____

_____

_____

_____

_____

3

KU | ES

Marks

1. (continued)

Diagram Q1C: A Holiday Home in the Lake District

Look at Diagrams Q1A and Q1C.

(c) There are plans to build twenty holiday homes in Area A (grid square 2805). Each home will look like the one shown in Diagram Q1C above.

Do you agree fully that these plans should go ahead?

Using map evidence, give reasons for your answer.

_____

_____

_____

_____

_____

_____

4

[Turn over

KU | ES

Marks

1. **(continued)**

(*d*) Do you think it will be possible for Ambleside (3704) to grow much further?

Give map evidence to support your answer.

_____

_____

_____

_____

_____

_____

4

(*e*) **"Brow Head Farm in grid square 3604 is likely to be a mixed farm."**

Do you agree fully with this statement?

Give reasons for your answer.

_____

_____

_____

_____

_____

3

KU | ES

Marks

KU	ES

Marks

1.   (continued)

Diagram Q1D: Single Carriageway Road   **Diagram Q1E: Dual Carriageway**

Look at Diagrams Q1A, Q1D and Q1E.

(*f*)   There is a plan to widen all of the A592 to a dual carriageway.

Why might people object to this?

Give map evidence to support your answer.

_____

_____

_____

_____

_____

_____   4

[Turn over

2.   **Diagram Q2: Synoptic Chart, 6 January, 0600 hours**

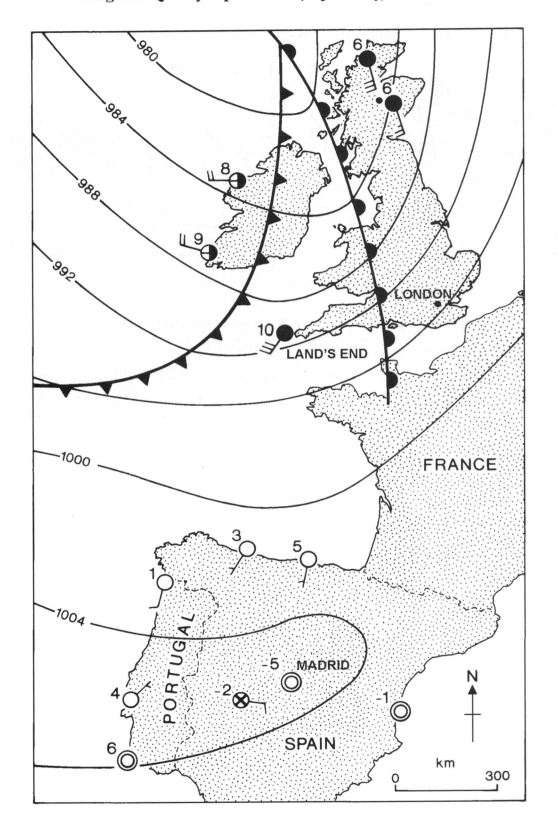

2. (continued)

Study Diagram Q2.

(a) Complete the station circle below to show the weather conditions at **London** on Diagram Q2.

Weather conditions at London

Wind from South West
Cloud cover: 4 oktas
Rain: Drizzle
Wind speed: 15 knots

8

3

(b) Study Diagram Q2.

Match the weather system to the locations given in the table.

Choose from: Anticyclone   Depression

Location	Weather System
British Isles	
Spain	

Give reasons for your answer.

_____

_____

_____

_____

_____   **4**

[Turn over

3.

Diagram Q3A: Selected Climate Regions

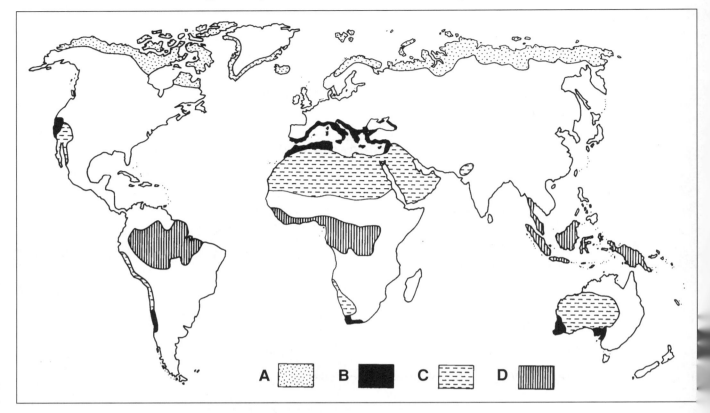

Diagram Q3B: Selected Climate Graphs

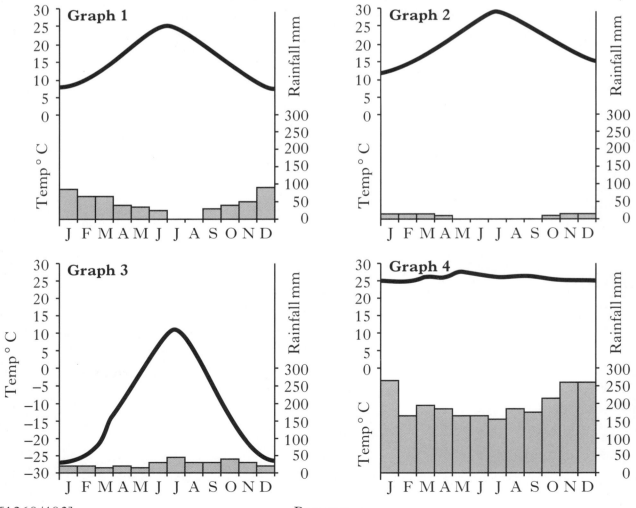

3. (continued)

Look at Diagrams Q3A and Q3B.

(a) Complete the table below by adding the appropriate letter or number.

Climate	Map Area	Graph
Hot desert		2
Equatorial Rainforest	D	
Mediterranean		1
Tundra	A	

2

(b) Describe, **in detail**, the main features of the climate shown in **Graph 4**.

_____

_____

_____

_____

_____

_____

3

[Turn over

4.   **Diagram Q4: New Developments at the Edge of Glasgow**

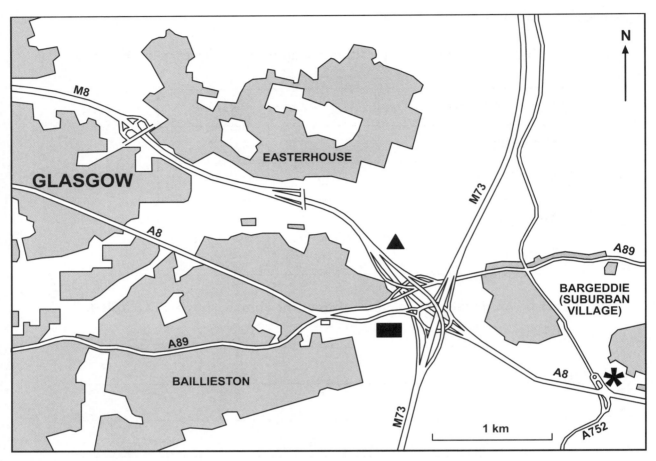

Key   Built-up area

New Developments

▲   Retail Park

■   Supermarket

✳   Leisure Complex

Marks

4. (continued)

Look at Diagram Q4.

(*a*) Select **one** of the developments shown and describe the advantages **and** disadvantages of its location.

Selected development _____

Advantages _____

_____

_____

Disadvantages _____

_____

_____ 4

(*b*) What **two** techniques could local pupils use to gather information about land use changes in the area shown in Diagram Q4?

Give reasons for your choices.

Technique 1 _____

Reason _____

_____

Technique 2 _____

Reason _____

_____ 4

[Turn over

5.   **Diagram Q5A: Aerial View in 2005 of Site for London Olympics**

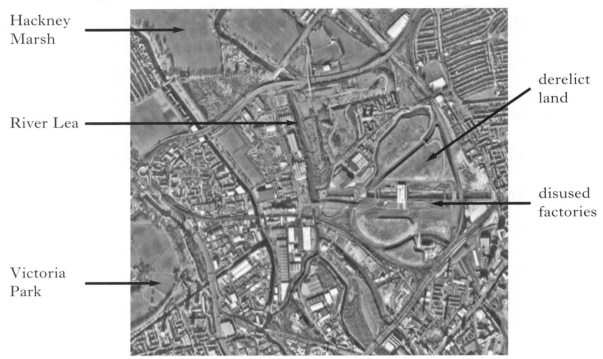

Hackney Marsh

derelict land

River Lea

disused factories

Victoria Park

Diagram Q5B: Plan of Olympic Site as proposed for 2012

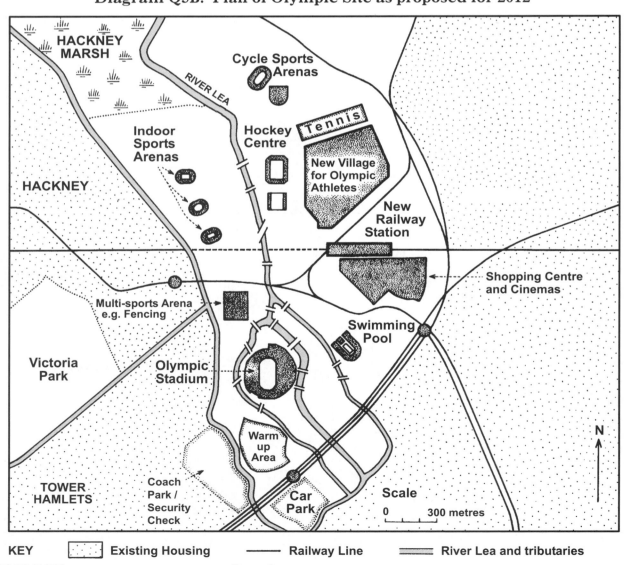

5. (continued)

Study Diagrams Q5A and Q5B.

The Olympic Games will be held in London in 2012.

How is this likely to benefit London?

_____

_____

_____

_____

_____

_____

4

[Turn over

6.   **Diagram Q6A: Braemore Farm, Northwest Scotland, in 1980**

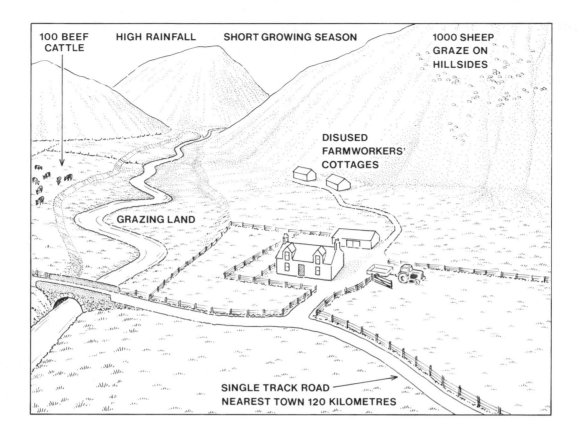

Diagram Q6B: Braemore Farm, Northwest Scotland, in 2010

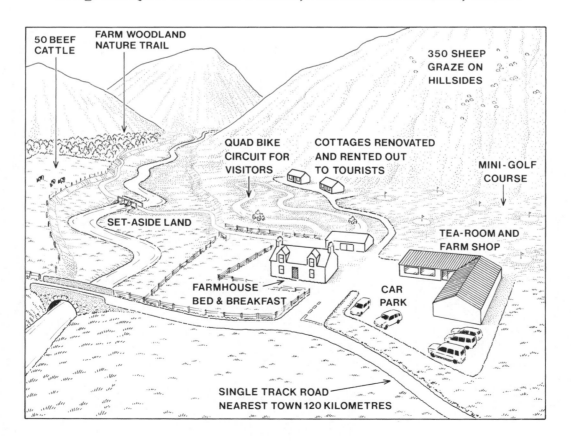

Marks

6.   (continued)

Study Diagrams Q6A and Q6B.

Braemore Farm in northwest Scotland has changed since 1980. Are these changes likely to have improved the farm?

Give reasons for your answer.

_____

_____

_____

_____

_____

_____   **4**

[Turn over

Marks

7. **Diagram Q7A: Age Structure of Population for Botswana (2000)**

Age Group	Males (% of population)	Females (% of population)
60+	3	3
20–59	20	23
0–19	25	26

Diagram Q7B: Population Pyramid for Botswana (2000)

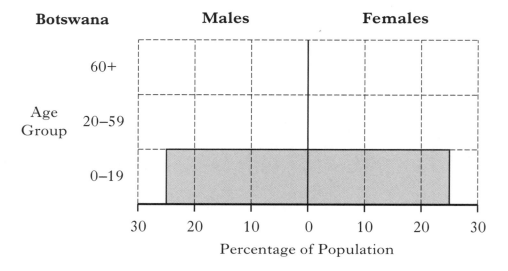

(a) Use the information shown in Diagram Q7A to complete the population pyramid for Botswana above (Diagram Q7B).

3

7.   **(continued)**

Diagram Q7C: Life Expectancy (2007)

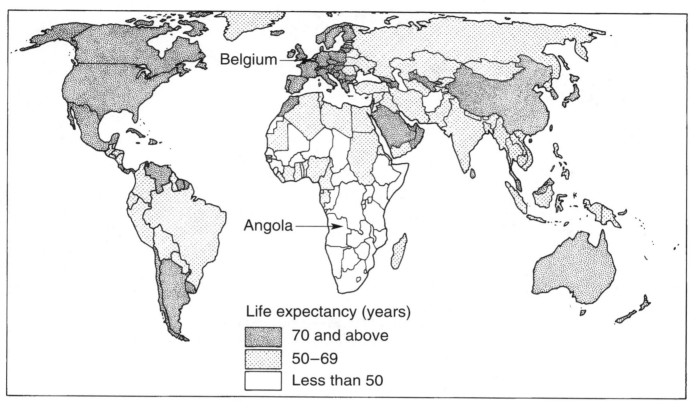

Belgium

Angola

Life expectancy (years)

70 and above

50–69

Less than 50

KU | ES

(*b*)   Look at Diagram Q7C above.

Explain why life expectancy is lower in Angola than in Belgium.

Marks

_____

_____

_____

_____

_____

_____

4

[Turn over

KU	ES

Marks

8.

Diagram Q8: Rural/Urban Population in Selected Countries

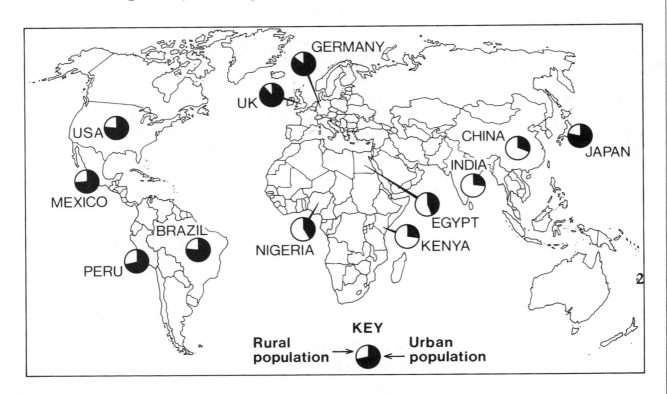

Look at Diagram Q8.

Name **two** other techniques which could be used to show the population information in Diagram Q8.

Give reasons for your choices.

Technique 1 _____

Reason(s) _____

_____

_____

Technique 2 _____

Reason(s) _____

_____

_____

4

DO NOT WRITE IN THIS MARGIN

KU | ES

Marks

9. Developing countries pay a tax (tariff) to sell their goods in EU countries. What problems can this cause for these developing countries?

_____

_____

_____

_____

_____

_____

3

[Turn over

10.

Diagram Q10A: Haiti

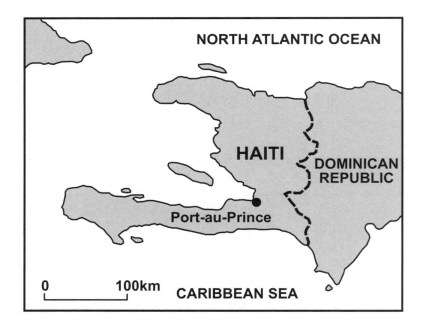

Diagram Q10B: Effects of Hurricanes in Haiti, August/September 2008

Deaths	500
Homeless	1 million
Crops destroyed	coffee, tobacco and sugar
Schools destroyed	30
Estimated cost of rebuilding	$180 million

Diagram Q10C: Types of Aid

Short-term Aid	Long-term Aid
Clean water	Rebuilding homes
Food	Road building
Emergency shelter	Electricity network
Medicines	Building hospitals

Marks

KU | ES

10. **(continued)**

Look at Diagrams Q10A, Q10B and Q10C.

In 2008, four hurricanes hit Haiti in twenty one days causing widespread destruction and flooding.

Which type of aid, **short-term** or **long-term**, would have been more useful to Haiti?

Explain your answer **in detail**.

_____

_____

_____

_____

_____

_____ **4**

[END OF QUESTION PAPER]

[BLANK PAGE]

STANDARD GRADE | CREDIT

2010

[BLANK PAGE]

C

1260/405

NATIONAL
QUALIFICATIONS
2010

THURSDAY, 6 MAY
1.00 PM – 3.00 PM

GEOGRAPHY
STANDARD GRADE
Credit Level

All questions should be attempted.

Candidates should read the questions carefully. Answers should be clearly expressed and relevant.

Credit will always be given for appropriate sketch-maps and diagrams.

Write legibly and neatly, and leave a space of about one centimetre between the lines.

All maps and diagrams in this paper have been printed in black only: no other colours have been used.

Extract No 1786/EXP367

1:25 000 Scale
Explorer Series

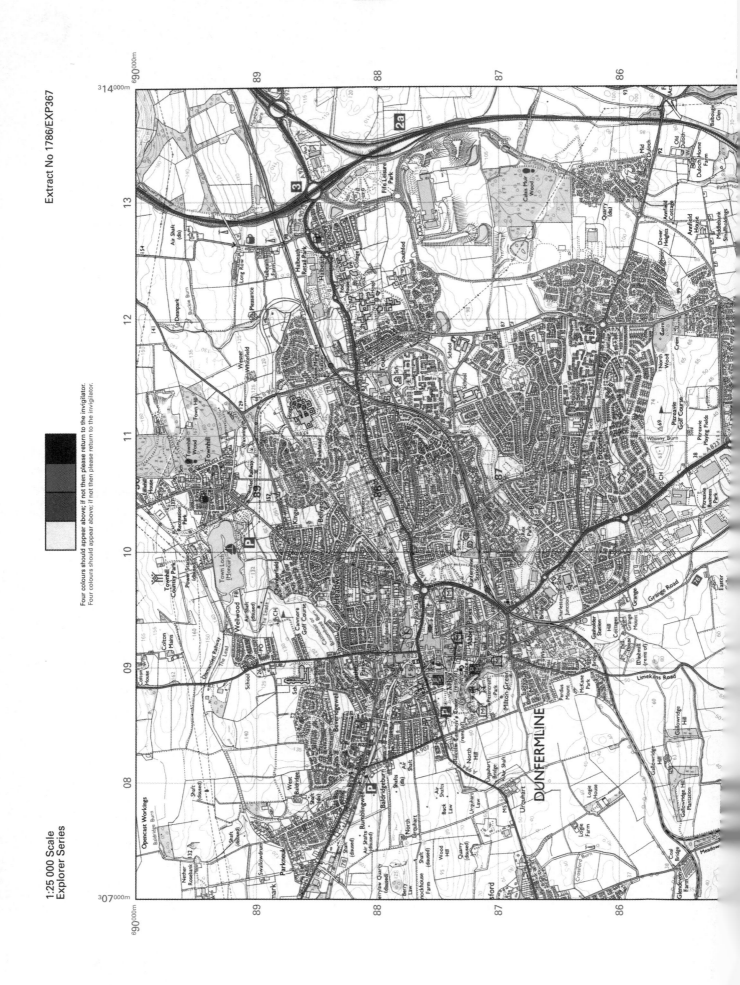

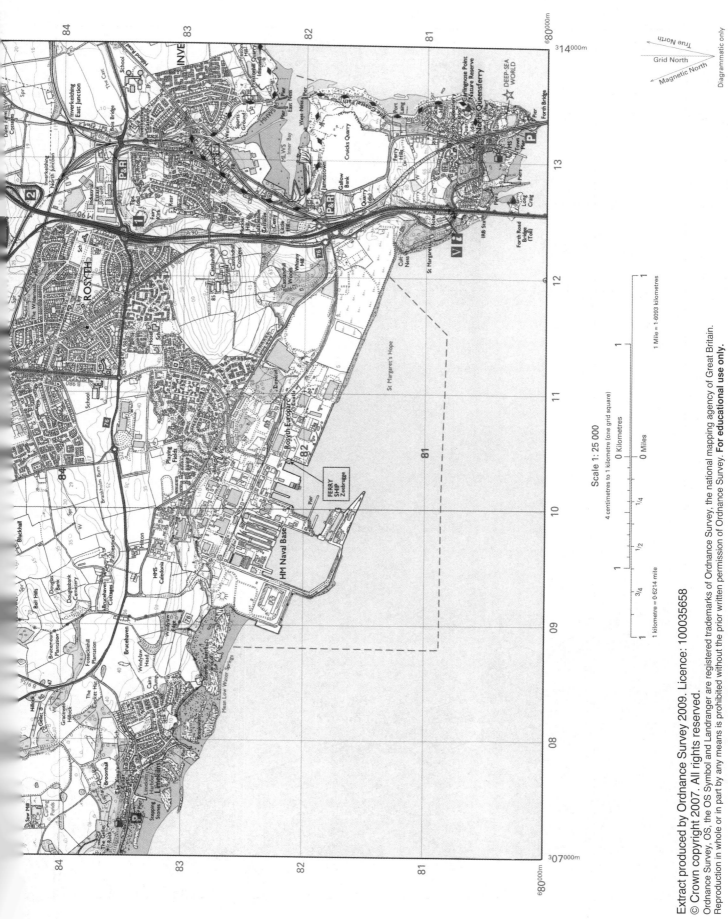

Scale 1 : 25 000

4 centimetres to 1 kilometre (one grid square)

1.

Diagram Q1

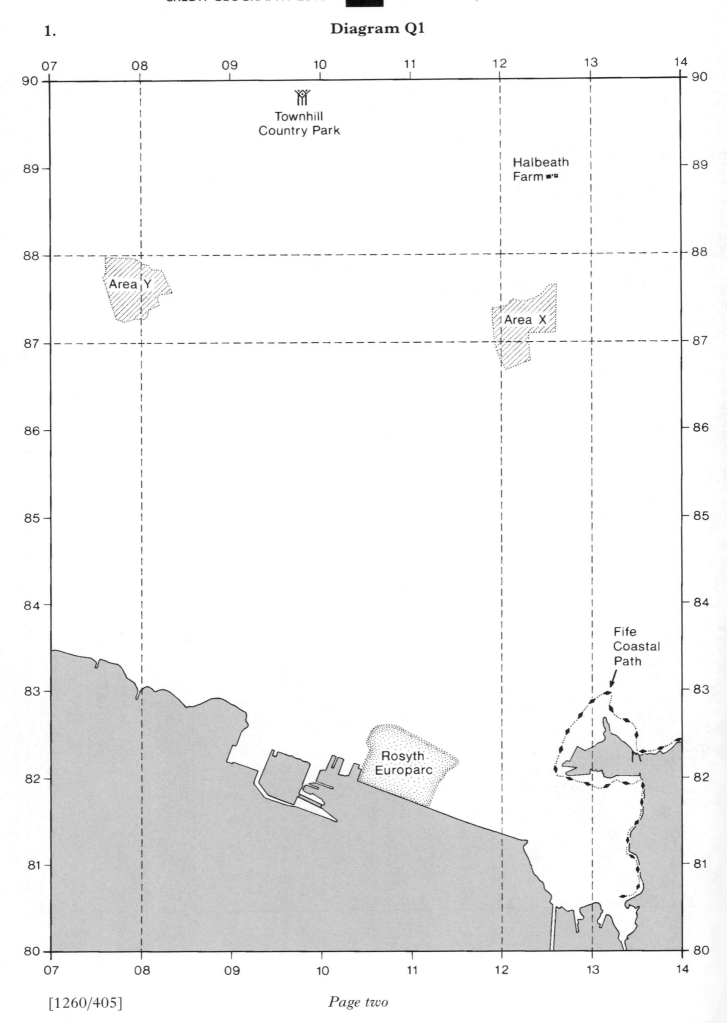

Marks

KU	ES

1. (continued)

This question refers to the OS Map (No 1786/EXP367) of the Dunfermline area and to Diagram Q1 on *Page two*.

(*a*) Find the route of the Fife Coastal Path between 134805 and 140824.

Describe the advantages **and** disadvantages of this route for walkers.

5

(*b*) What techniques could a group of Geography students use to gather information about Townhill Country Park (0989)?

Give reasons for your choice of techniques.

5

(*c*) What is the main present day function of Dunfermline?

Choose from: service centre   tourist centre.

Use map evidence to support your answer.

4

(*d*) There is a plan to build a new housing estate at either Area X (around 120870) or Area Y (around 082877).

Which location is better?

Give reasons for your choice.

5

(*e*) **"Halbeath Farm at 126889 is an excellent location for farming."**

Do you agree fully with this statement?

Give reasons for your answer.

4

(*f*) Rosyth Europarc (110820) is an industrial estate.

Explain the advantages of its location.

4

[Turn over

Marks
KU | ES

2.   **Diagram Q2A: A Meander**   **Diagram Q2B:**
                                         Cross Section of a Meander

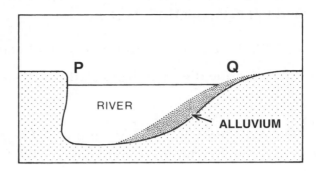

Look at Diagrams Q2A and Q2B.

Explain the different processes at work in the river at points P and Q.

4

3.   **Diagram Q3: An Anticyclone over Europe**

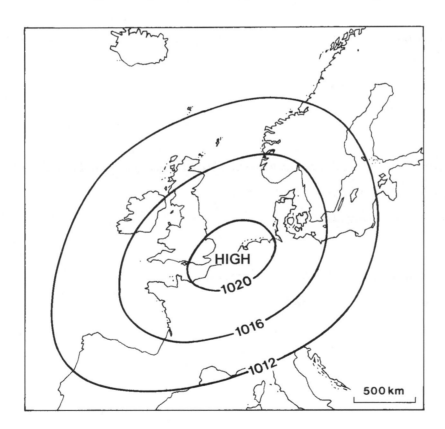

Look at Diagram Q3.

Describe similarities **and** differences in weather conditions caused by anticyclones in summer and winter.

4

Marks

KU | ES

4.

Diagram Q4A:	Diagram Q4B:
The Sahel Zone, Africa	**Population (millions) in Sahel Countries**

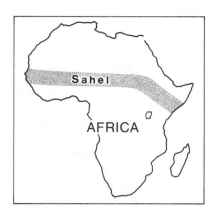

Country	Year		
	1950	**1995**	**2005**
Ethiopia	20	52	67
Sudan	9	30	35
Chad	3	6	9

Diagram Q4C: Causes of Desertification

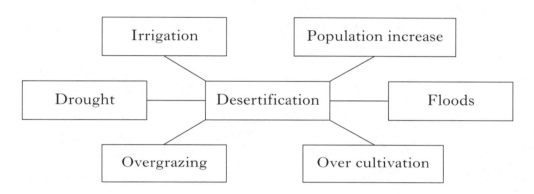

Look at Diagrams Q4A, Q4B and Q4C.

"Population is the main cause of desertification."

Do you agree fully with this statement?

Give reasons for your answer.

5

[Turn over

5.   **Diagram Q5: Development at Menie Estate**

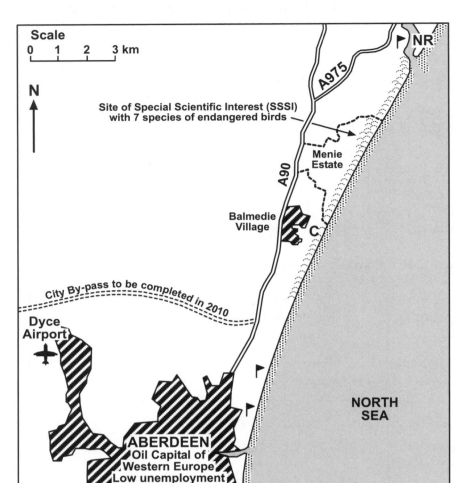

Look at Diagram Q5.

A foreign businessman is developing the **Menie Estate** as a luxury golf resort.

What are the advantages **and** disadvantages of this development for the area?

Marks

KU | ES

6.   **Diagram Q6A: Land Use Zones in Urban Areas**

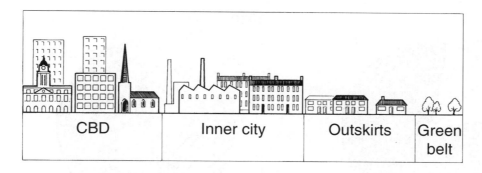

CBD | Inner city | Outskirts | Green belt

Diagram Q6B: Changing Land Values from Centre to Edge of City

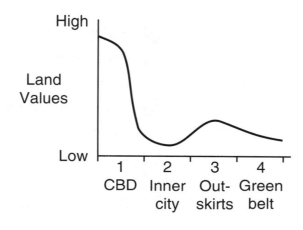

Look at Diagrams Q6A and Q6B.

Choose any **two** zones shown in Diagram Q6A. **For each zone chosen**, give reasons for the ways in which the land has been used.

6

[Turn over

7. **Diagram Q7: Percentage Income at Redland Farm**

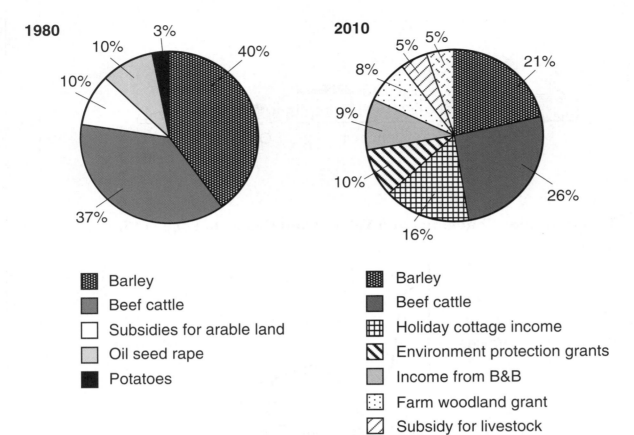

1980

3%
10%
10%
40%
37%

	Barley
	Beef cattle
	Subsidies for arable land
	Oil seed rape
	Potatoes

2010

5% 5%
5%
8%
9%
21%
10%
16%
26%

	Barley
	Beef cattle
	Holiday cottage income
	Environment protection grants
	Income from B&B
	Farm woodland grant
	Subsidy for livestock
	Organic potatoes

Look at Diagram Q7.

The sources of this farmer's income have changed since 1980.

Give reasons for these changes.

6

8.

| Diagram Q8A: | Diagram Q8B: |
| Kenya, Population Density | Kenya, Annual Rainfall |

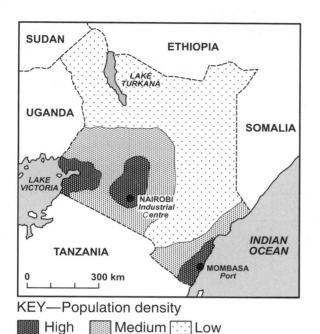

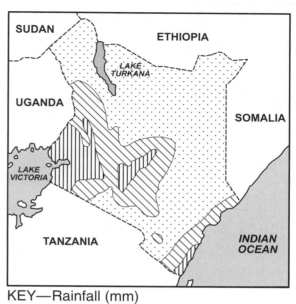

KEY—Population density

■ High ▦ Medium ⬚ Low

KEY—Rainfall (mm)

▥ More than 1000   ◩ 500–1000   ⬚ Less than 500

Diagram Q8C: Agriculture and Industry

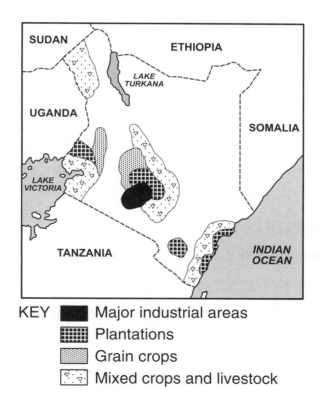

KEY ■ Major industrial areas
     ▦ Plantations
     ▨ Grain crops
     ⬚ Mixed crops and livestock

Study Diagrams Q8A, Q8B and Q8C.

Explain the distribution of population in Kenya.

4

9.   **Diagram Q9A: Population Growth in India**

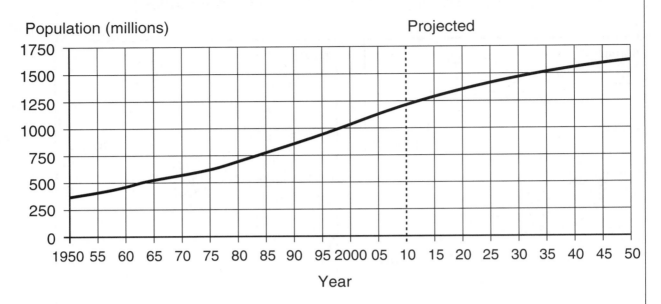

Diagram Q9B: Factors linked to Population Increase

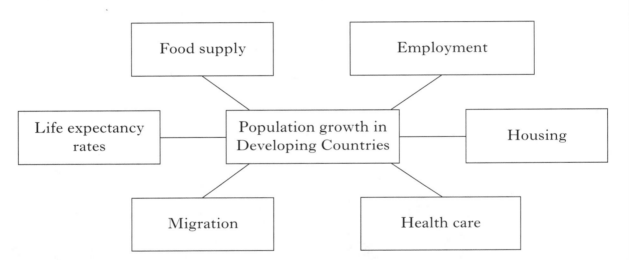

Look at Diagrams Q9A and Q9B.

"The population growth of India will cause the country many problems."

Do you agree fully with this statement?

Give reasons for your answer.

Marks
KU | ES

10. **Diagram Q10: Selected Developing Countries where one raw material is more than half of all exports**

Zambia Copper 87%

Bangladesh Jute 51%

Cuba Sugar 77%

Laos Timber 76%

Ghana Cocoa 80%

Look at Diagram Q10.

(*a*) What problems does this pattern of trade cause in developing countries? **4**

(*b*) Which **other** processing techniques could be used to display the export figures shown on Diagram Q10?

Give reasons for your choice of techniques. **5**

[Turn over for Question 11 on *Page twelve*

11. **Diagram Q11: Some Charities providing Aid to Developing Countries**

Look at Diagram Q11.

In what ways is aid given by charities (voluntary aid) suitable for developing countries?

4

[END OF QUESTION PAPER]

STANDARD GRADE | GENERAL

2011

[BLANK PAGE]

FOR OFFICIAL USE

G

KU	ES

Total Marks

1260/403

NATIONAL
QUALIFICATIONS
2011

TUESDAY, 24 MAY
10.25 AM–11.50 AM

GEOGRAPHY
STANDARD GRADE
General Level

Fill in these boxes and read what is printed below.

Full name of centre

Town

Forename(s)

Surname

Date of birth

Day   Month   Year

Scottish candidate number

Number of seat

1 Read the whole of each question carefully before you answer it.

2 Write in the spaces provided.

3 Where boxes like this ☐ are provided, put a tick ✓ in the box beside the answer you think is correct.

4 Try all the questions.

5 Do not give up the first time you get stuck: you may be able to answer later questions.

6 Extra paper may be obtained from the Invigilator, if required.

7 Before leaving the examination room you must give this book to the Invigilator. If you do not, you may lose all the marks for this paper.

Extract No 1879/EXP308

1:25 000 Scale
Explorer Series

Four colours should appear above; if not then please return to the invigilator.
Four colours should appear above; if not then please return to the invigilator.

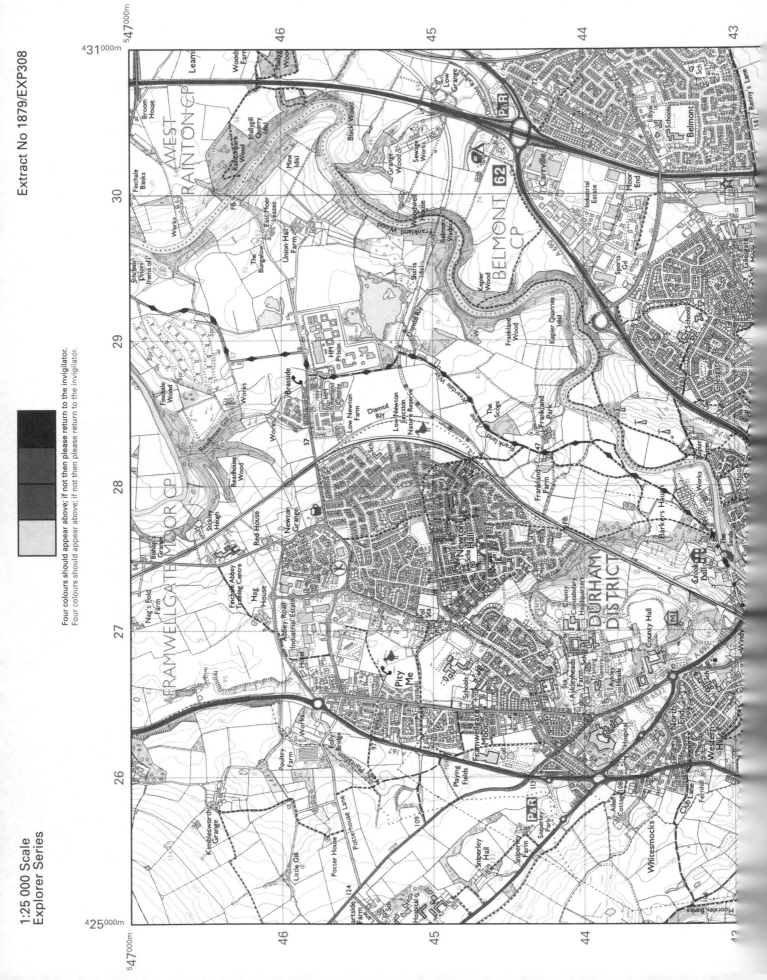

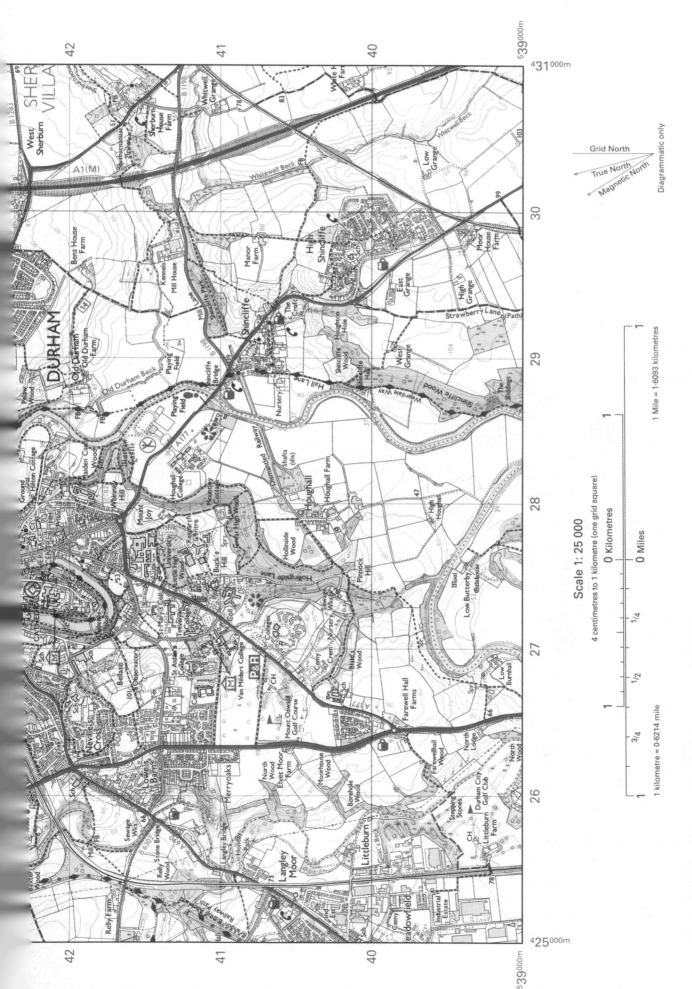

Scale 1: 25 000

4 centimetres to 1 kilometre (one grid square)

1.

Diagram Q1A

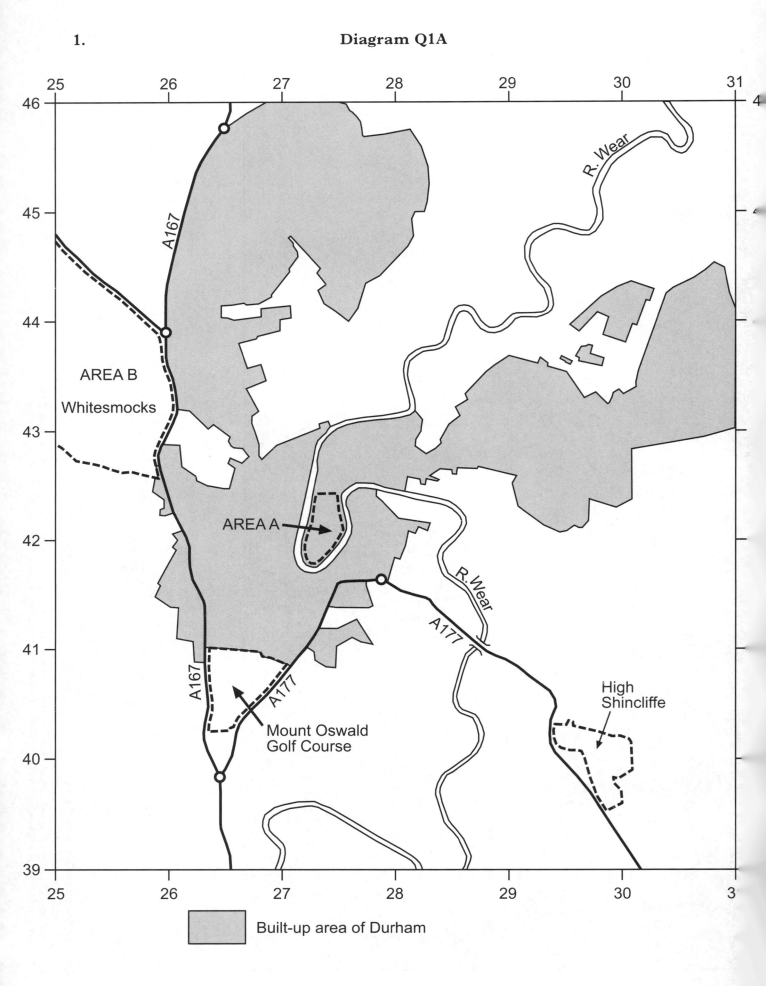

Built-up area of Durham

Marks | KU | ES

1. **(continued)**

Look at the Ordnance Survey Map Extract (No 1879/308) and Diagram Q1A on *Page two*.

(*a*) Describe the **physical** features of the River Wear from 287390 to 280432.

_____

_____

_____

_____

_____

3

(*b*) Look at Diagram Q1A and the Map Extract.

The original site of Durham is in Area A on Diagram Q1A.

Giving map evidence, suggest what features attracted the earliest settlers to this site.

_____

_____

_____

_____

_____

3

(*c*) Find Area B on Diagram Q1A and on the Map Extract.

"It would be difficult for Durham to expand into Area B."

Do you agree fully with this statement?

Give reasons for your answer.

_____

_____

_____

_____

_____

_____

4

Marks KU ES

1. (continued)

(*d*) Find High Shincliffe on Diagram Q1A and on the Map Extract.

"A dormitory settlement is a community where most of the residents travel to a larger settlement to work."

Give map evidence to show that High Shincliffe is a dormitory settlement.

_____

_____

_____

_____

_____

3

1.   (continued)

Diagram Q1B: Mount Oswald Development Scheme

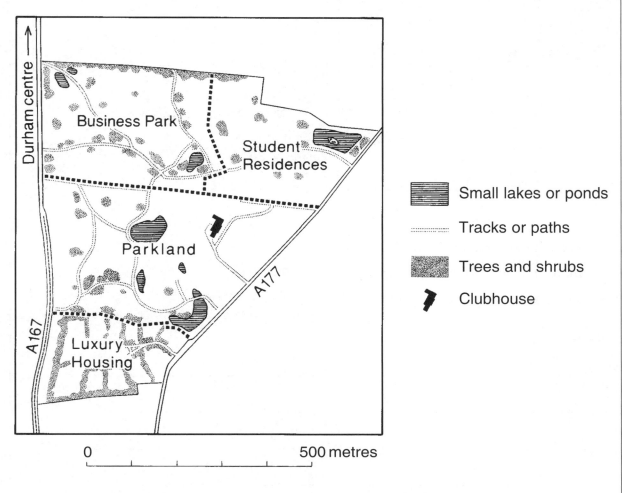

(e)   Study the Map Extract, Diagram Q1A and Diagram Q1B.

There are plans to change Mount Oswald golf course into a business park, residential areas and parkland.

"This would be a good site for the new development."

Do you agree fully with this statement?

Give reasons for your answer.

_____

_____

_____

_____

_____

_____

4

2.   **Diagrams Q2A and Q2B: Sketches of a River's Course**

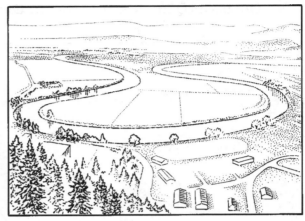

A   B

Look at Diagrams Q2A and Q2B.

Which Diagram shows the upper course of the river?

Give reasons for your choice.

Choice: Diagram _____

Reasons _____

_____

_____

_____

_____

_____   4

Marks

3.　**Diagram Q3:　Effects of Snowfall in UK, February 2009**

| Icy pavements |

| Heathrow, Gatwick and London City Airports closed |

| Increased sales of sledges, salt, gloves and wellingtons |

| Six million people take the day off |

| Ski resorts report business up by 20% |

| Business loses over £1·2 billion |

Look at Diagram Q3.

The snowfall in February 2009 was the biggest for 18 years.

"This caused only problems."

Do you agree fully with this statement?

Give reasons for your answer.

_____

_____

_____

_____

_____

_____　4

[Turn over

4.
Diagram Q4A: Climate Stations 1, 2, 3 and 4

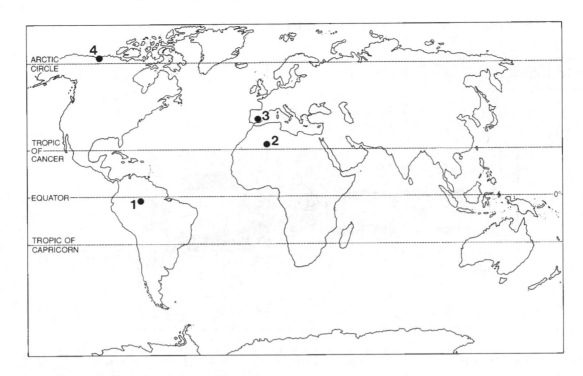

Diagram Q4B: Climate Graphs

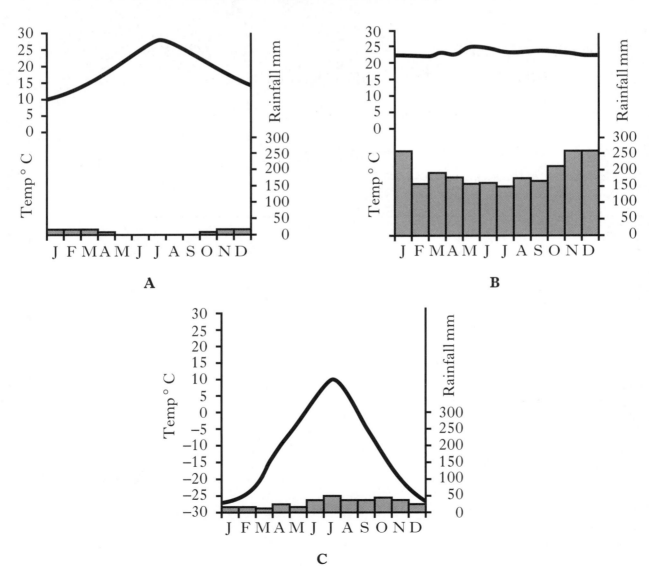

Marks | KU | ES

4. (continued)

(a) Look at Diagrams Q4A and Q4B.

Complete the table below by matching the correct climate stations 1, 2, 3 or 4 to the correct graph.

Climate Graph	Climate Station
A	
B	
C	

3

(b) Choose **one** of the climate graphs shown in Diagram Q4B.

Tick (✓) your choice.

A ☐   B ☐   C ☐

Describe how this climate affects the way people live.

_____

_____

_____

_____

_____

3

[Turn over

Marks   KU   E

5. **Diagram Q5: Causes of Deforestation in the Amazon Basin**

Cause of Deforestation	Percentage (%)
A Subsistence agriculture	30
B Cattle ranching	60
C Logging	7
D Mining, roads, dams and towns	3

(*a*) Look at Diagram Q5.

Complete the pie chart below using the figures in Diagram Q5 and label it correctly.

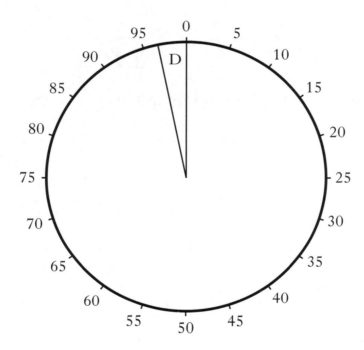

3

5. (continued)

(*b*) Choose **one** of the following and **explain** its impact on the environment of the Amazon Basin.

logging   subsistence agriculture   cattle ranching.

_____

_____

_____

_____

_____

_____

3

[Turn over

Marks KU ES

6.   **Diagram Q6: Newspaper Headlines on Inner City Developments**

New expensive apartments to replace office blocks in city centre

Playing fields to be sold for supermarket development

Tenements to be pulled down for new office block

Old disused factory to be replaced by a leisure centre

Old church buildings to be used as pubs and clubs

"None of the developments above should be allowed to go ahead."

Local resident

Do you agree completely with the statement above?

Give reasons for your answer.

_____

_____

_____

_____

_____

_____

4

Marks | KU | ES

7.   **Diagram Q7: Factors affecting Glen Rinnes Farm, Morayshire**

| long distance footpath | narrow strips of woodland | gently sloping land | steep hilly land |

| nearest town 6 miles: Dufftown (population 1450) | access via B9009 | 310 metres above sea level |

Study Diagram Q7.

"Glen Rinnes Farm is an excellent location for a farm."

Do you agree fully with this statement?

Give reasons for your answer.

_____

_____

_____

_____

_____

4

8.   **Diagram Q8A: Former Paper Mill at Inverurie**

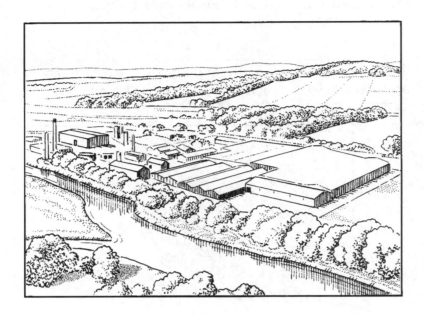

Diagram Q8B: Location of former Inverurie Paper Mill

Inverurie:
Population 11,000

Former Inverurie
Paper Mill

A96 to Aberdeen
Airport (10 miles)
Aberdeen (15 miles)

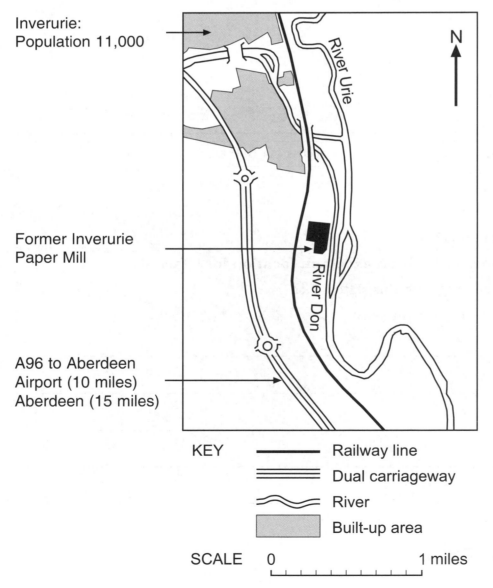

KEY ——————— Railway line

Dual carriageway

River

Built-up area

SCALE   0   1 miles

Marks | KU | ES

8. (continued)

(*a*) Study Diagrams Q8A and Q8B.

Inverurie Paper Mill closed down in 2009.

A group of geography pupils wish to find out about the effects of the closure of the mill.

What gathering techniques could they use?

Give reasons for your choices.

_____

_____

_____

_____

_____

_____ **4**

(*b*) Many companies have expressed an interest in the site of Inverurie Paper Mill.

Explain the advantages of this site **for** manufacturing industry.

_____

_____

_____

_____

_____

_____ **4**

[Turn over

Marks KU E

9.   Diagram Q9A: Changing Age Structure, Italy 2000–2025

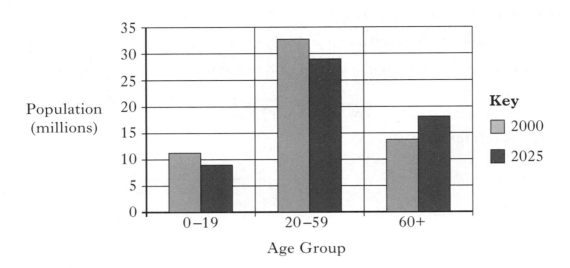

(*a*) Study Diagram Q9A.

Describe the changes in Italy's age structure between 2000 and that projected for 2025.

_____

_____

_____

_____

_____

_____   3

9.   (continued)

Diagram Q9B: Maps of Italy

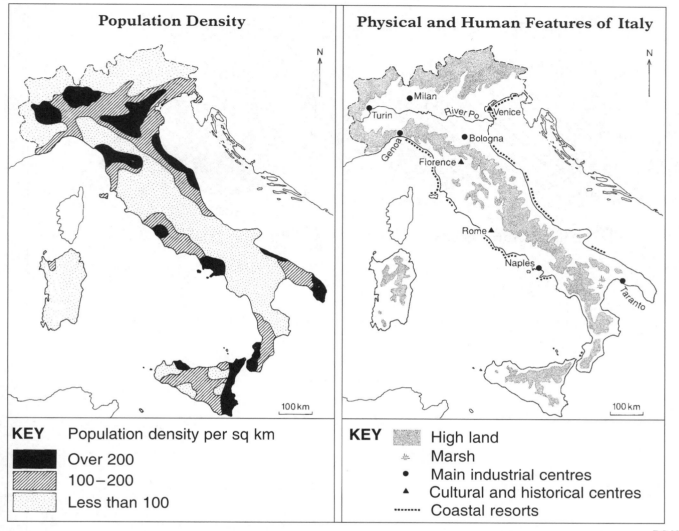

Population Density

Physical and Human Features of Italy

KEY   Population density per sq km

- Over 200
- 100–200
- Less than 100

KEY
- High land
- Marsh
- Main industrial centres
- Cultural and historical centres
- Coastal resorts

Marks | KU | ES

(b)   Study Diagram Q9B.

What are the links between population density and the physical and human features of Italy?

4

Marks KU

10. Diagram Q10: Exports of Developed and Developing Countries

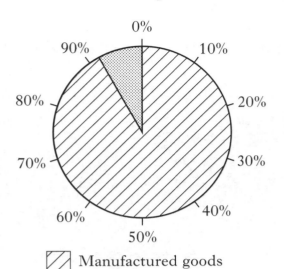

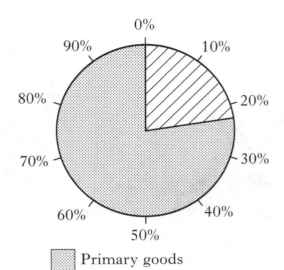

Look at Diagram Q10.

(*a*) How does the trade pattern shown in Diagram Q10 cause problems for developing countries?

_____

_____

_____

_____

_____   **3**

(*b*) Name **two** other techniques which could be used to show the information in Diagram Q10.

Give reasons for your choice.

Technique 1 _____

Reason _____

_____

_____

Technique 2 _____

Reason _____

_____

_____   **4**

11.   **Diagram Q11: Features of Self-Help Schemes**

Use simple tools and equipment		Use materials available locally
	SELF-HELP SCHEMES	
Employ many local people	Low cost	Set up in rural communities

Explain how Self-Help Schemes are suitable for Developing Countries.

_____

_____

_____

_____   **3**

[END OF QUESTION PAPER]

[BLANK PAGE]

STANDARD GRADE | CREDIT

2011

[BLANK PAGE]

C

1260/405

NATIONAL
QUALIFICATIONS
2011

TUESDAY, 24 MAY
1.00 PM – 3.00 PM

GEOGRAPHY
STANDARD GRADE
Credit Level

All questions should be attempted.

Candidates should read the questions carefully. Answers should be clearly expressed and relevant.

Credit will always be given for appropriate sketch-maps and diagrams.

Write legibly and neatly, and leave a space of about one centimetre between the lines.

All maps and diagrams in this paper have been printed in black only: no other colours have been used.

PB 1260/405 6/15910

Extract No 1880/20

1:50 000 Scale
Landranger Series

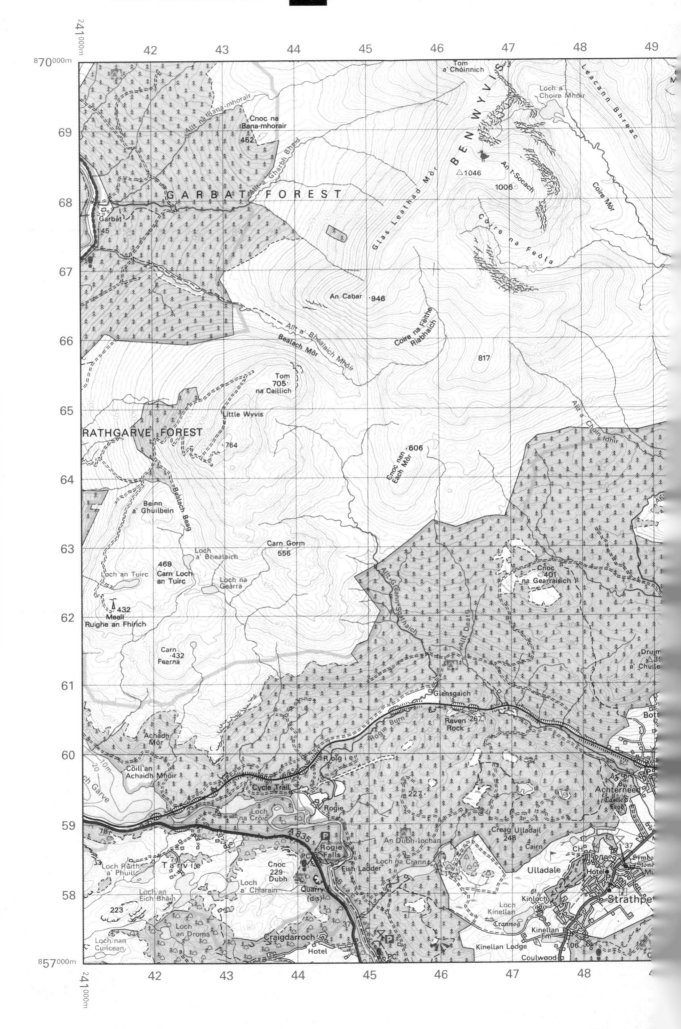

Grid North

Magnetic North

True North

Diagrammatic only

Scale 1:50 000

2 centimetres to 1 kilometre (one grid square)

1 mile = 1·6093 kilometres

1 kilometre = 0·6214 mile

1. **Diagram Q1A**

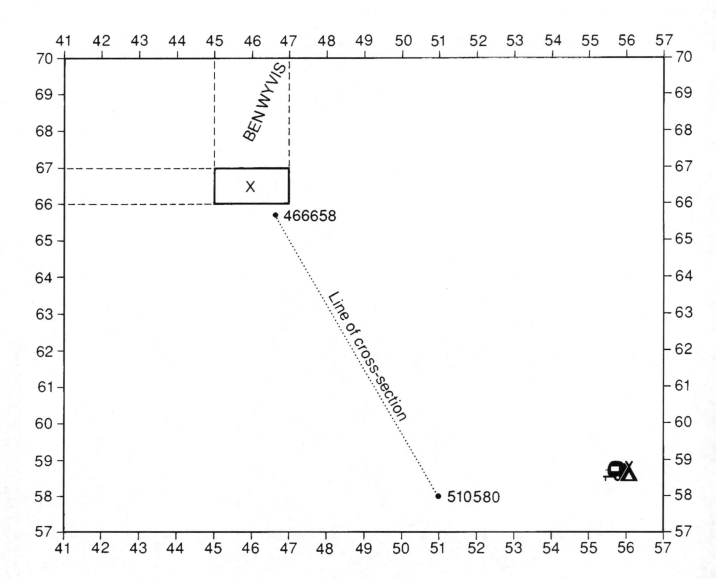

Marks
KU | ES

1. (continued)

This question refers to the OS Map (No 1880/20) of the Dingwall area and to Diagram Q1A on *Page two*.

(*a*) Match each of the features named below with the correct grid reference.

Features: hanging valley, corrie, truncated spur.

Choose from grid references: 525658, 467677, 435663, 476683.

3

Diagram Q1B: Cross-section from GR 466658 to GR 510580 showing Land Use

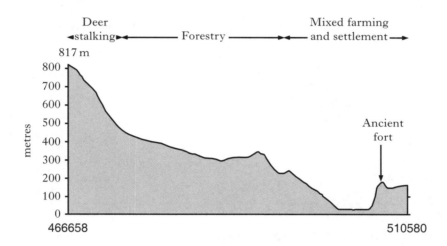

(*b*) Look at the map extract and find the cross-section shown on Diagrams Q1A and Q1B.

Study both the map extract and the cross-section.

Give reasons for the differences in land use along the section.

5

[Turn over

Marks
KU | ES

1.   **(continued)**

(c) Look at the map extract and find Area X marked on Diagram Q1A.

There is a proposal to develop this area as a ski centre.

What are the advantages **and** disadvantages of this location for such a development?

Give map evidence to support your answer.

5

(d)   **Diagram Q1C: Desirable Qualities for National Park Status**

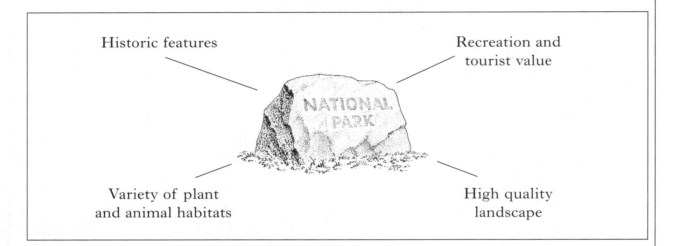

Look at the map extract and Diagram Q1C above.

To what extent do you agree that the whole map area would be suitable for a National Park?

You must refer to at least **two** qualities shown in Diagram Q1C.

Give map evidence to support your answer.

(e) Find the caravan and camping site in grid square 5558.

Is this a good location for a caravan and camping site?

Give map evidence to support your answer.

2.   **Diagram Q2: Carn Mor Dearg Arete**

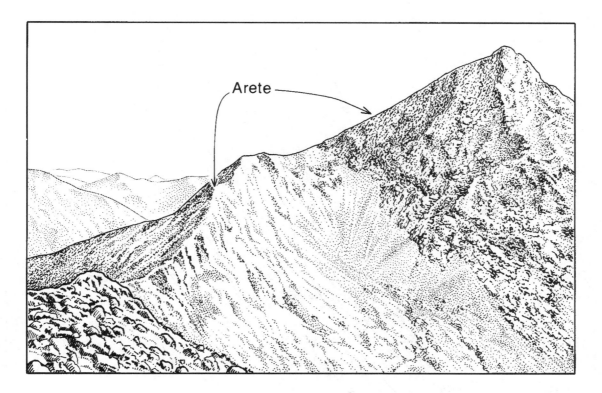

Look at Diagram Q2.

Explain how the arete was formed.

You may use a diagram(s) to illustrate your answer.

4

[Turn over

3.   **Diagram Q3A: Weather Chart, 31 December 2006, 4.00 pm**

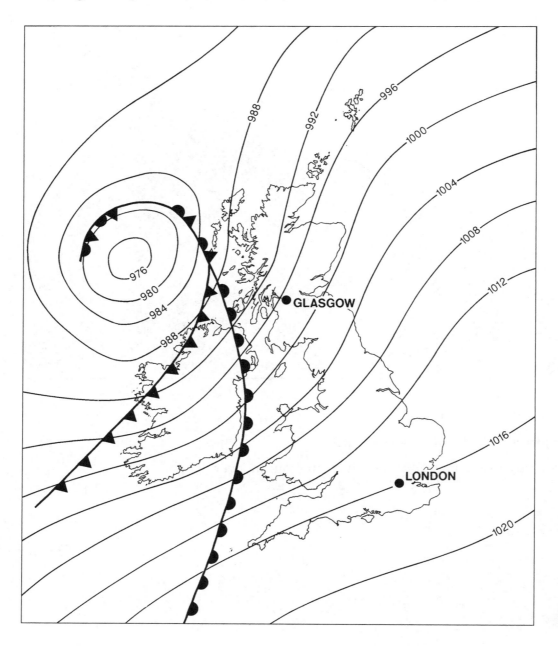

Diagram Q3B: Newspaper Headlines, 1 January 2007

Glasgow cancels Hogmanay Street Party just a few hours before midnight.

Celebrations went ahead in London where thousands brought in the New Year in Trafalgar Square.

Study Diagrams Q3A and Q3B.

With reference to the weather chart, **explain** why London's New Year celebrations went ahead whilst the celebrations in Glasgow were cancelled.

5

4.   **Diagram Q4A: Palm Oil Production and Rainforest Cover 1964–2009**

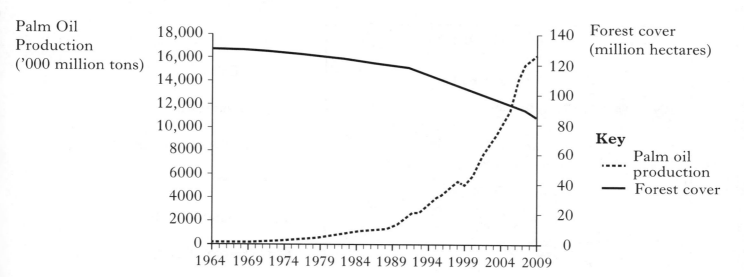

Palm Oil Production ('000 million tons)

Forest cover (million hectares)

Key

- - - - Palm oil production
───── Forest cover

Diagram Q4B: Some Facts about Palm Oil and Indonesia

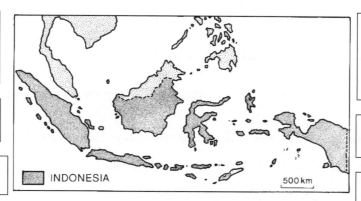

Habitat of Indonesian Orangutans (apes) replaced by 4·3 million hectares of oil palm plantations

75% of greenhouse gases are a result of deforestation

Palm Oil provides money for amenities such as clean water, electricity and schools

Rainforest home to native people

Average earnings are $3900 per annum

International debt of $155 billion

INDONESIA   500 km

Rainforests are the world's "lungs"

World demand for Palm Oil is predicted to double by 2030

Palm Oil production employs over 3·5 million

Look at Diagrams Q4A and Q4B.

"The destruction of the rainforest is a small price to pay for the huge benefits it has brought to Indonesia."

To what extent do you agree with this statement?

Give reasons for your answer.

Marks

KU	ES
	5

[Turn over

5.   **Diagram Q5: Different Zones in a City**

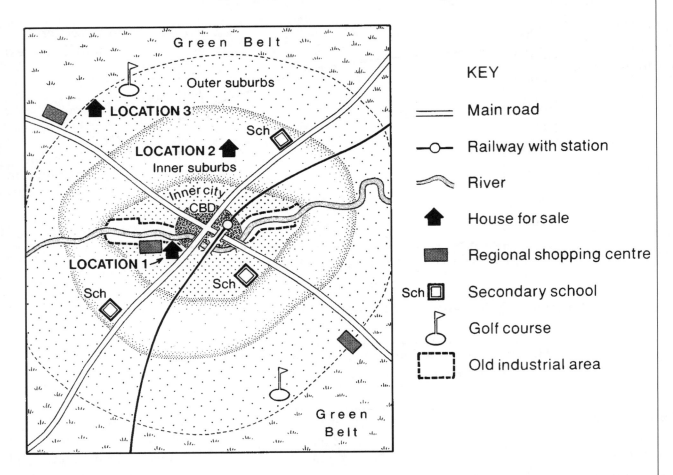

(a) Mr and Mrs Boal and their two teenage children are moving to the city shown above.

They have found houses for sale in the three locations identified.

Which location do you think they should choose?

Give reasons for your choice.

(b) Describe techniques which pupils could use to gather information about the differences between the urban environments of the three locations.

Give reasons for your choice of techniques.

Marks
KU | ES

6.   **Diagram Q6: Recent Trends in Farming**

| Farm Shop Sales Hit Record High |

| EU Slashes Farm Subsidies |

| Consumers Turn to Organic Produce |

| Foot and Mouth Restrictions Hit Farmers Hard |

| Global Warming Causes Rise in Insect Pests |

| Open Access Policy Encourages More People into the Countryside |

| Environment Stewardship Grants For Farms up 60% |

| Fuel Prices Rocket |

Look at Diagram Q6.

"Prospects for Scottish farmers in 2012 look bleak and uncertain."

Do you agree fully with this statement?

Give reasons for your answer.

5

[*Turn over*

7.

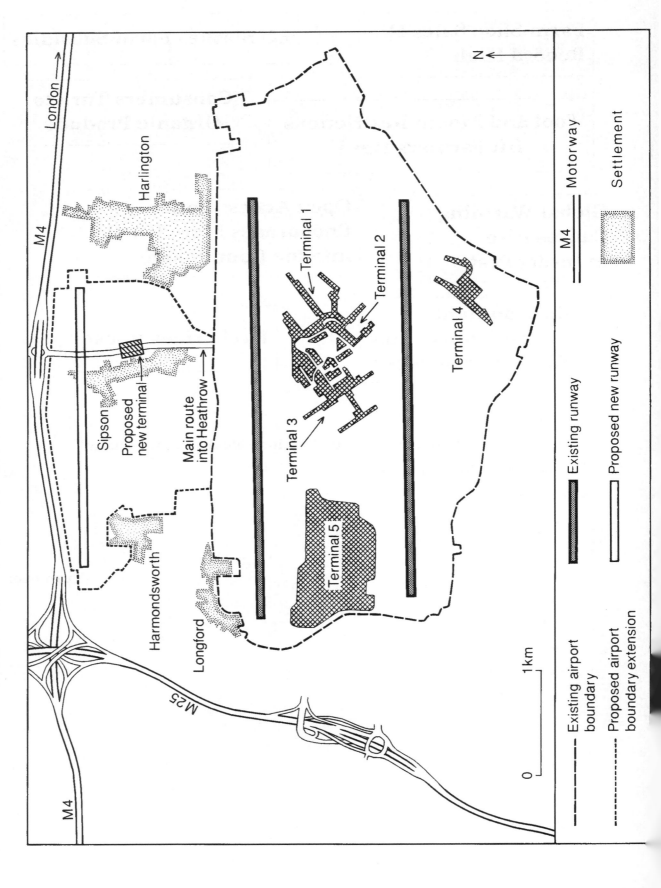

Diagram Q7A: Proposed Expansion of London's Heathrow Airport

7. **(continued)**

Diagram Q7B: Arguments for and against the Expansion of Heathrow

Study Diagrams Q7A and Q7B.

What are the advantages **and** disadvantages of the expansion of Heathrow Airport for South East England?

5

[Turn over

8.   **Diagram Q8: Heriot Watt Research Park, Edinburgh**

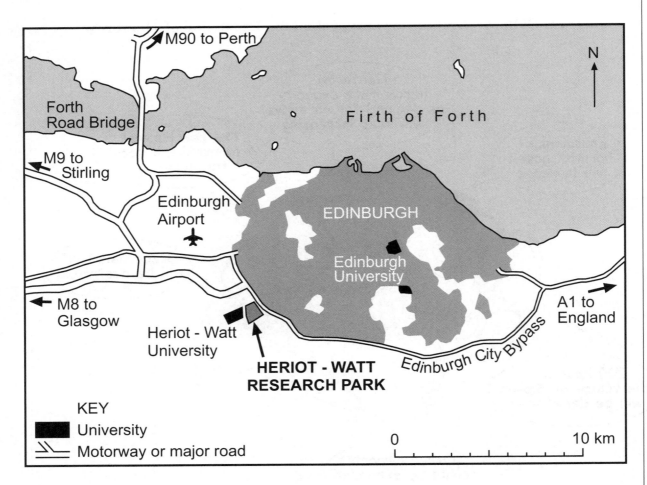

Look at Diagram Q8.

A research park is an **industrial area** where companies research **and** manufacture high-tech goods such as micro-electronics equipment and bio-technology products.

Explain why Heriot Watt Research Park is a good location for companies.

5

Marks

| KU | ES |

9.

Diagram Q9A: Location of Kenya, Africa

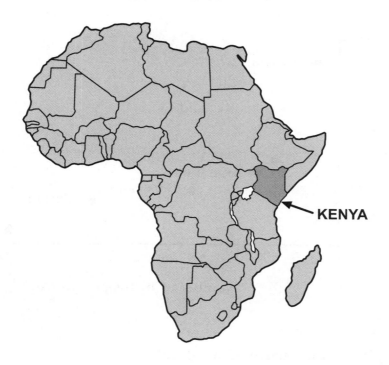

KENYA

Diagram Q9B: Projected Population Statistics for Kenya

Year	2000	2025	2050
% living in rural areas	80	70	52
% living in urban areas	20	30	48

(a) Look at Diagrams Q9A and Q9B above.

Give reasons for the population changes in Kenya.   **5**

(b) Look at Diagram Q9B above.

What other techniques could be used to show the information in the table?

Give reasons for your choice.   **5**

[Turn over

10.   **Diagram Q10: GDP per Capita and Number of Births per Woman
in selected Countries**

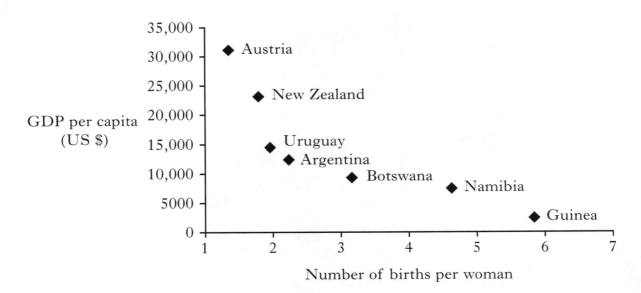

Look at Diagram Q10 above.

"GDP per capita is a measure of wealth."

Explain why there is a link between GDP per capita and the number of births per woman as shown in the graph.

5

11.   **Diagram Q11A: Selected Population Data**

EU (including UK)	490 million
USA	305 million
Japan	127 million
UK	60 million

Diagram Q11B: Selected Features of the EU (European Union)

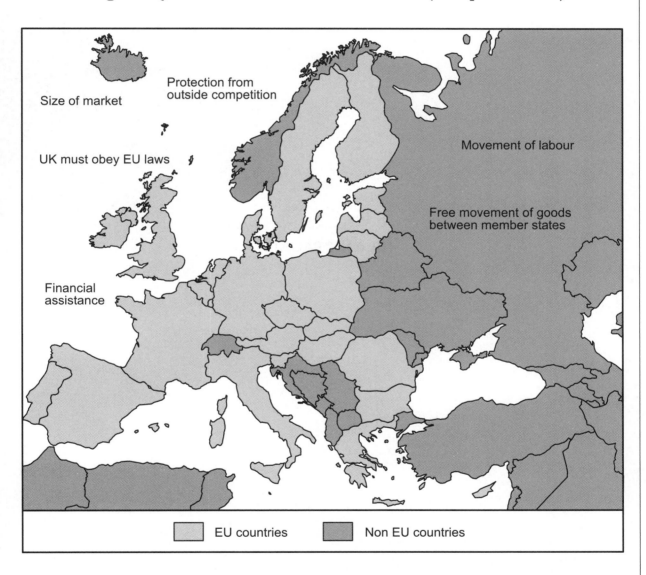

Look at Diagrams Q11A and Q11B.

"For the UK, membership of the EU brings more advantages than disadvantages."

Do you agree fully with this statement?

Give reasons for your answer.

[*END OF QUESTION PAPER*]

4

[BLANK PAGE]

STANDARD GRADE | GENERAL

2012

[BLANK PAGE]

FOR OFFICIAL USE

G

KU	ES

Total Marks

1260/29/01

NATIONAL
QUALIFICATIONS
2012

TUESDAY, 8 MAY
10.25 AM – 11.50 AM

**GEOGRAPHY
STANDARD GRADE**
General Level

Fill in these boxes and read what is printed below.

Full name of centre

Town

Forename(s)

Surname

Date of birth

Day	Month	Year	Scottish candidate number	Number of seat

1   Read the whole of each question carefully before you answer it.

2   Write in the spaces provided.

3   Where boxes like this ☐ are provided, put a tick ✓ in the box beside the answer you think is correct.

4   Try all the questions.

5   Do not give up the first time you get stuck: you may be able to answer later questions.

6   Extra paper may be obtained from the Invigilator, if required.

7   Before leaving the examination room you must give this book to the Invigilator. If you do not, you may lose all the marks for this paper.

1:50 000 Scale
Landranger Series

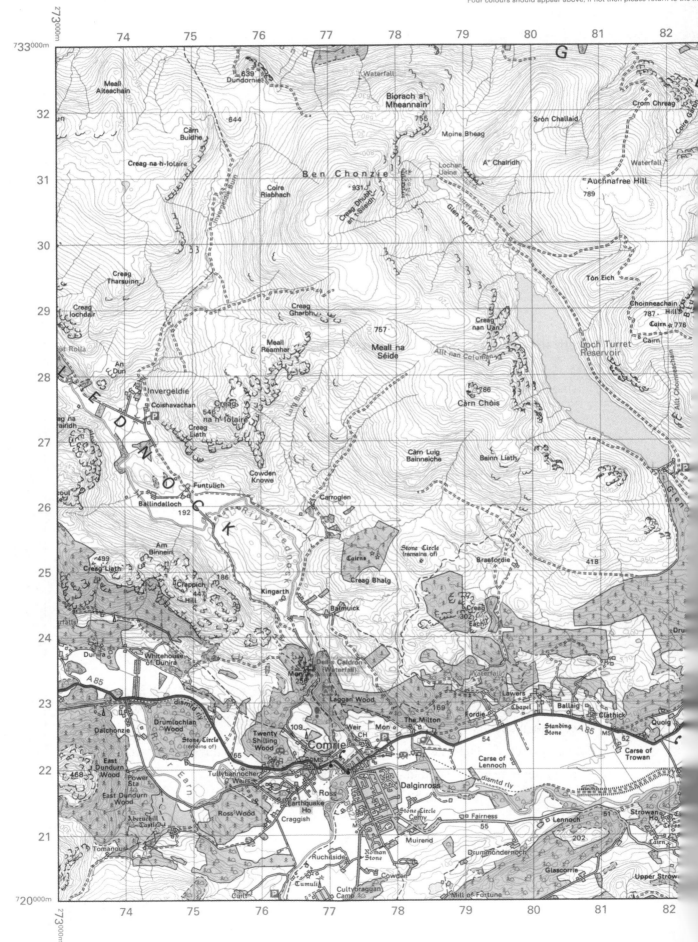

Extract No 1937/52

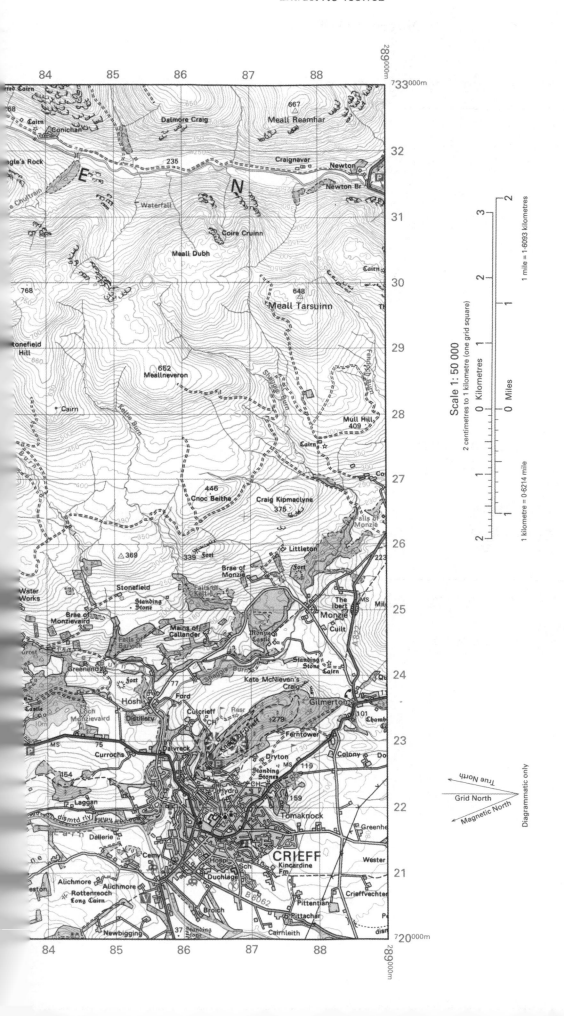

Scale 1: 50 000

2 centimetres to 1 kilometre (one grid square)

1 kilometre = 0·6214 mile

1 mile = 1·6093 kilometres

True North

Grid North

Magnetic North

Diagrammatic only

1. **Diagram Q1A**

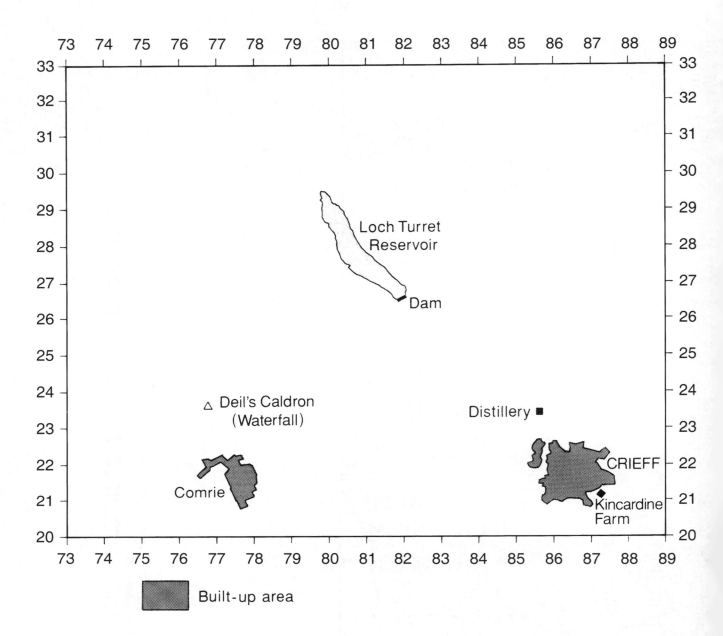

Marks | KU | ES

1. (continued)

Look at the Ordnance Survey Map Extract (No 1937/52) of the Crieff area and Diagram Q1A on *Page two*.

(*a*) (i) Match each of the grid references below with the correct feature in the table. Choose from:

8029   8732   7830   8731.

Landscape Feature	Grid Reference
Corrie loch	
U-shaped valley	
Hanging valley	
Truncated spur with crags	

3

(ii) **Explain** how a **U-shaped valley** is formed.

You may use diagrams to illustrate your answer.

_____

_____

_____

_____

_____

_____

_____

_____

_____

_____

3

Marks   KU   ES

1. **(continued)**

(b) Find Kincardine Farm at grid reference 8721.

What are the advantages and disadvantages of this site for a farm?

Advantages _____

_____

_____

_____

Disadvantages _____

_____

_____

_____   **4**

(c) What is the main function of Crieff?

Tick (✔) your choice. Give reasons for your answer.

Choose from:

Market town   ☐

Tourist centre   ☐

Reasons _____

_____

_____

_____   **4**

Marks   KU   ES

1.   (continued)

(d)   **Diagram Q1B: Whisky Distillery Inputs**

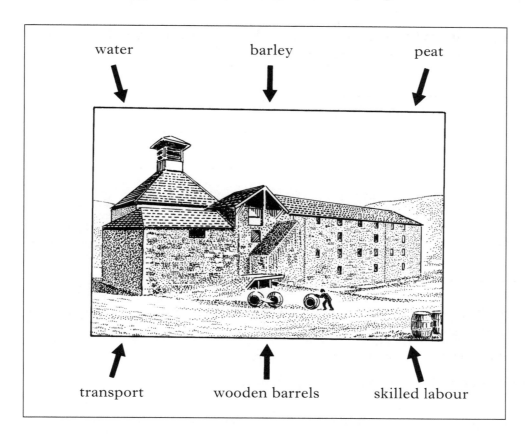

There is a whisky distillery at 8523.

Is this a good place for a whisky distillery?

Using map evidence, give reasons for your answer.

_____

_____

_____

_____

_____

_____   **4**

[Turn over

Marks   KU | ES

1. (continued)

(*e*) The Deil's Caldron Waterfall (7623) attracts thousands of visitors to the Comrie area each year.

Using map evidence, describe, in detail, the problems for the area caused by this number of visitors.

4

(*f*) Loch Turret Reservoir was formed by building a dam in grid squares 8126 and 8226.

Explain why this area is suitable for a reservoir.

You **must** refer to map evidence in your answer.

3

Marks | KU | ES

2.   **Diagram Q2: A Glaciated Lowland Landscape**

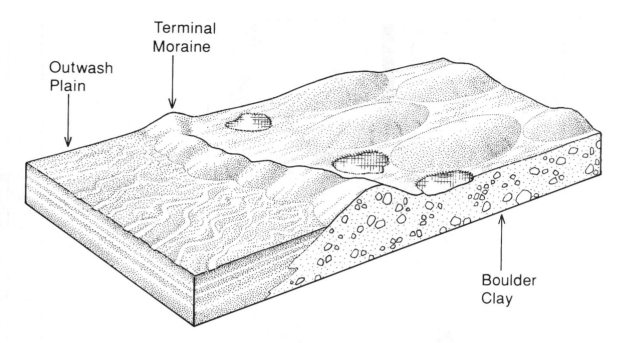

Look at Diagram Q2 above.

Do you agree fully that farming is the best way to use this area?

Give reasons for your answer.

_____

_____

_____

_____   **3**

[Turn over

DO NO
WRIT
IN TH
MARG

Marks KU

3.   **Diagram Q3: Air Masses affecting the British Isles**

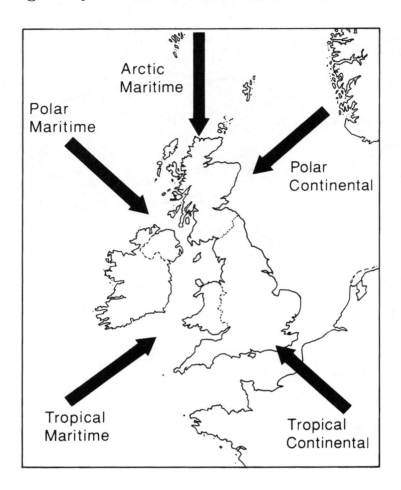

Study Diagram Q3 above.

Describe the problems which an Arctic Maritime air mass is likely to bring to northern Scotland in winter.

_____

_____

_____

_____

_____   3

Marks | KU | ES

4.   **Diagram Q4A: Climate Data for Iraklion, Greece**

	J	F	M	A	M	J	J	A	S	O	N	D
Rainfall (mm)	97	85	28	28	11	0	1	3	14	35	63	111
Temperature (°C)	12	12	14	16	20	23	26	26	24	21	17	15

Diagram Q4B: Climate Graph for Iraklion, Greece

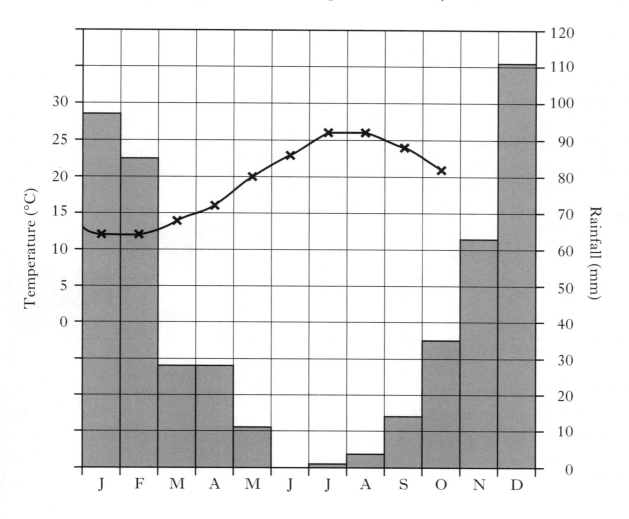

(a)   Look at the climate data for Iraklion (Diagram Q4A).

Use this data to complete the climate graph (Diagram Q4B).   2

Complete part (a) above before answering part (b) below.

(b)   Describe, in detail, the climate of Iraklion.

_____

_____

_____

_____

_____

_____   3

5.   **Diagram Q5: Solutions to Desertification**

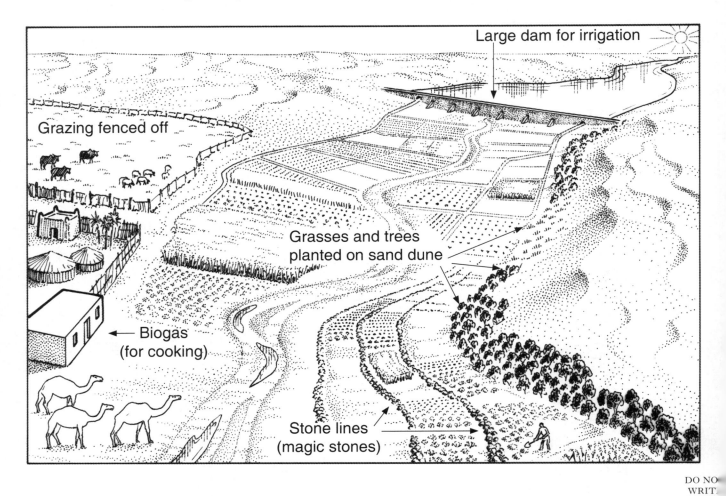

Large dam for irrigation

Grazing fenced off

Grasses and trees
planted on sand dune

Biogas
(for cooking)

Stone lines
(magic stones)

Study Diagram Q5 above.

Choose **two** solutions shown in Diagram Q5.

Explain how they help to reduce desertification in **developing countries**.

Solution 1 _____

_____

_____

_____

Solution 2 _____

_____

_____

_____

4

Marks KU

DO NO
WRIT
IN TH
MARGI

6.   **Diagram Q6: Land Use Zones in Perth**

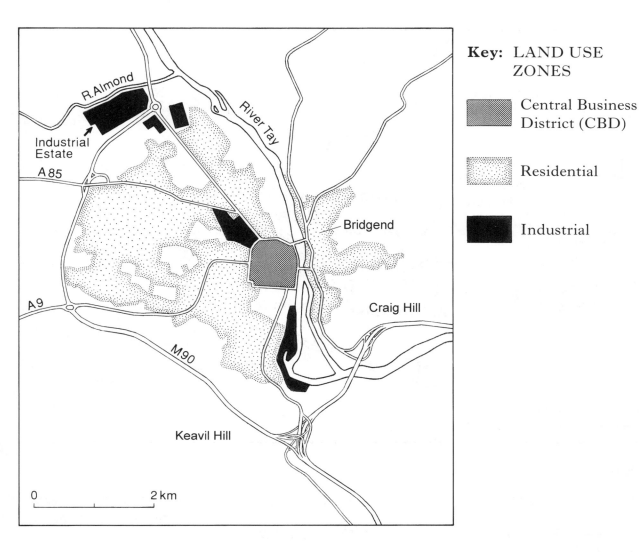

Key: LAND USE ZONES

Central Business District (CBD)

Residential

Industrial

Study Diagram Q6 above.

Choose **one** land use zone and **explain** its location.

Choice _____

_____

_____

_____

_____

_____

_____   4

7.   **Diagram Q7: Industrial Developments west of London**

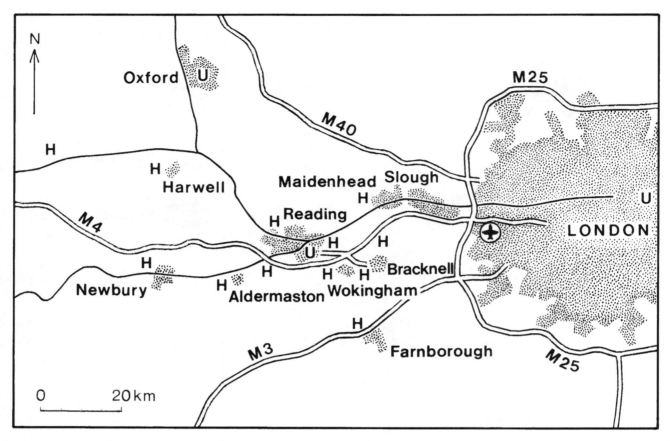

Key

	Motorway	**U**	University	**H**	High-tech firms and research centres
	Railway	✈	Heathrow airport		
	Towns and cities				

7. (continued)

(a) Look at Diagram Q7.

Explain why so many high-tech firms and research centres are located in the area shown.

_____

_____

_____

_____

_____

_____   **4**

(b) What gathering techniques could be used to find out about the effects of these industrial developments on the area?

Give reasons for your choice of techniques.

Techniques _____

_____

_____

Reasons _____

_____

_____

_____   **4**

[Turn over

Marks KU ES

8.   **Diagram Q8: Hedgerows in the countryside**

Scenery

Wildlife

Space

Costs

Shelter

Look at Diagram Q8 above.

"Since 1945, 25% of UK hedgerows have been removed from farmland."

Describe the advantages and disadvantages of this change.

Advantages _____

_____

_____

_____

Disadvantages _____

_____

_____

_____   4

Marks KU ES

9. **Diagram Q9: World Population Distribution**

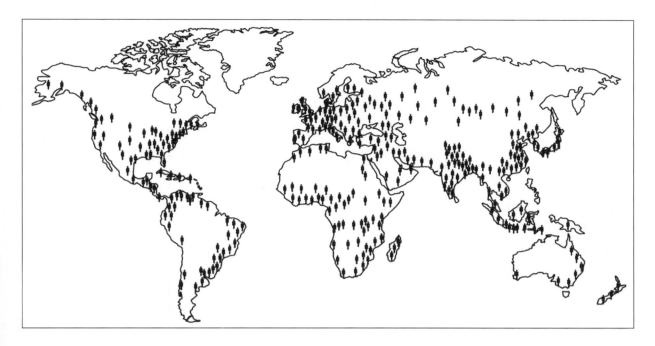

Look at Diagram Q9.

Explain why the population of the world is not evenly distributed.

_____

_____

_____

_____

_____

_____ 3

[Turn over

Marks KU E

10.   **Diagram Q10: Average Number of Children per Family in the UK,
1880–2010**

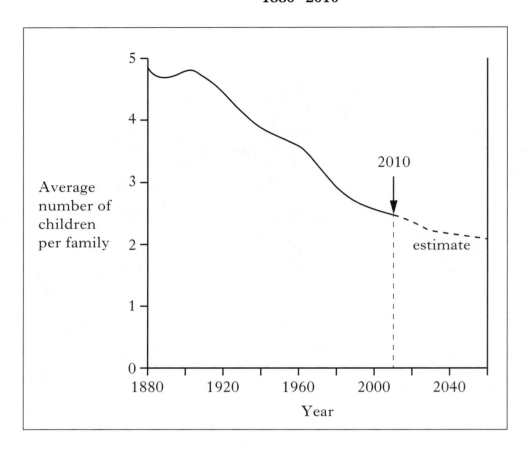

Look at Diagram Q10.

**"The average number of children per family in the UK has dropped
from five to two."**

Explain why people in developed countries such as the UK are having
smaller families.

_____

_____

_____

_____

_____   3

11. **Diagram Q11: Pattern of Trade for a Developing Country**

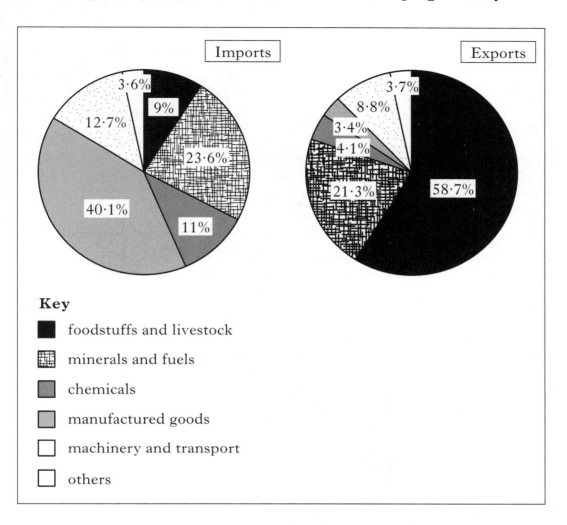

Give **two other** processing techniques which could be used to show the information in Diagram Q11.

Give reasons for your choice of techniques.

Technique 1   _____

Reason _____

_____

Technique 2   _____

Reason _____

_____

4

[Turn over

12.   **Diagram Q12A: Floods in Pakistan, Summer 2010**

Diagram Q12B: Effects of Floods in Pakistan, Summer 2010

Deaths	2000
Injured or homeless	21 million
Houses destroyed or damaged	1·9 million
Area flooded	160,000 km^2
Estimated economic cost	$43 billion

Marks

DO NOT WRITE IN THIS MARGIN

KU   ES

12. (continued)

Diagram Q12C: Types of Aid Required

Short-term Aid	Long-term Aid
Medical supplies	Rebuilding homes
Food and clean water	Road reconstruction
Tents and blankets	Help for farmers
Helicopters	New hospitals and schools

Look at Diagrams Q12A, Q12B and Q12C.

In summer 2010 large areas of Pakistan were badly affected by flooding.

Which type of aid, **short-term** or **long-term** would have been more useful to Pakistan?

Give reasons for your choice.

Choice _____

Reasons _____

_____

_____

_____

_____

_____

4

[END OF QUESTION PAPER]

[BLANK PAGE]

STANDARD GRADE | CREDIT

2012

[BLANK PAGE]

C

1260/31/01

NATIONAL
QUALIFICATIONS
2012

TUESDAY, 8 MAY
1.00 PM – 3.00 PM

GEOGRAPHY
STANDARD GRADE
Credit Level

All questions should be attempted.

Candidates should read the questions carefully. Answers should be clearly expressed and relevant.

Credit will always be given for appropriate sketch-maps and diagrams.

Write legibly and neatly, and leave a space of about one centimetre between the lines.

All maps and diagrams in this paper have been printed in black only: no other colours have been used.

1:50 000 Scale
Landranger Series

Extract No 1938/102

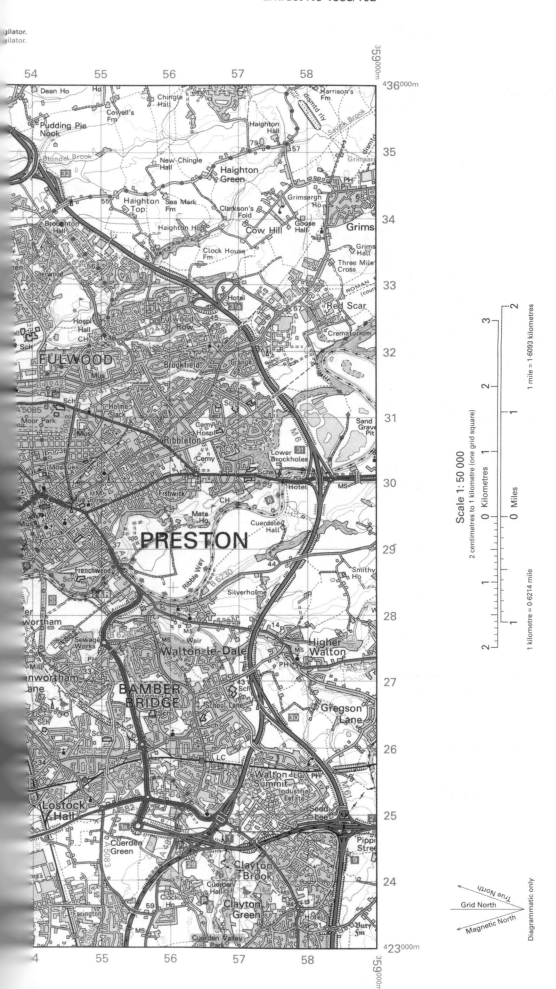

Scale 1: 50 000

2 centimetres to 1 kilometre (one grid square)

True North
Grid North
Magnetic North

Diagrammatic only

1.

Diagram Q1A

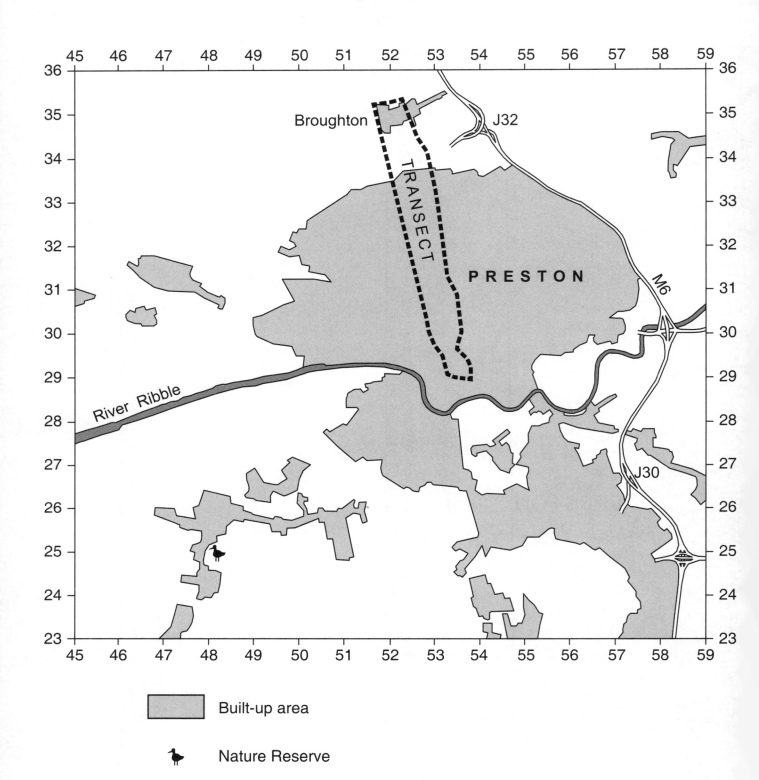

Marks

KU | ES

1. (continued)

This question refers to the OS Map (No 1938/102) of the Preston area and to Diagram Q1A on *Page two*.

(*a*) Describe the **physical** features of the River Ribble **and** its valley from 590305 to 450277.

4

(*b*) There is an industrial estate at 5725 (Walton Summit).

What are the advantages **and** disadvantages of this location for an industrial estate?

Use map evidence to support your answer.

5

(*c*) There is a nature reserve around 480250.

Is this a suitable site for a nature reserve?

Give reasons for your answer.

4

(*d*) Find the M6 east of Preston.

How did road engineers overcome the problems caused by the landscape between Junction 30 (5726) and Junction 32 (5434)?

4

[Turn over

Marks

KU | ES

1. **(continued)**

Diagram Q1B: Land Use Transect

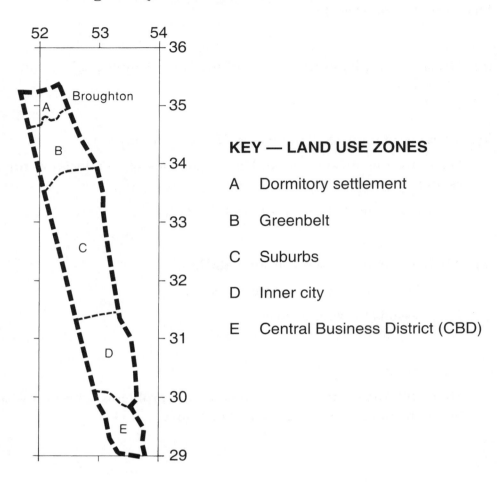

KEY — LAND USE ZONES

A   Dormitory settlement

B   Greenbelt

C   Suburbs

D   Inner city

E   Central Business District (CBD)

(*e*)   A group of school pupils has carried out fieldwork along the transect shown on Diagrams Q1A and Q1B.

They have correctly identified the land use zones shown on Diagram Q1B.

Referring to at least **three** zones, give map evidence to support the pupils' findings.

6

(*f*)   **"A dormitory settlement is a community where most of the residents travel to work in a larger settlement."**

What gathering techniques might the pupils have used to prove that Broughton is a dormitory settlement?

Give reasons why these techniques would have been appropriate.

Marks

KU | ES

2.   **Diagram Q2: The Formation of a Corrie**

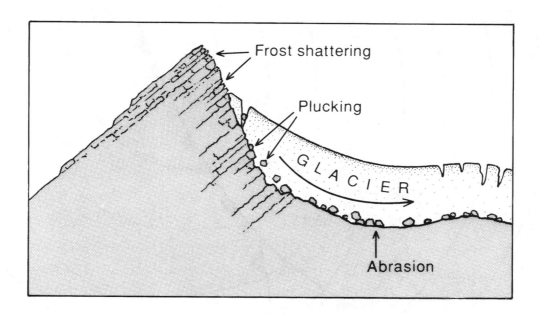

Look at Diagram Q2.

Explain, in detail, how the processes shown in the diagram produce a corrie.

5

[Turn over

3.   **Diagram Q3: Synoptic Chart, 0800 hours, 10 March**

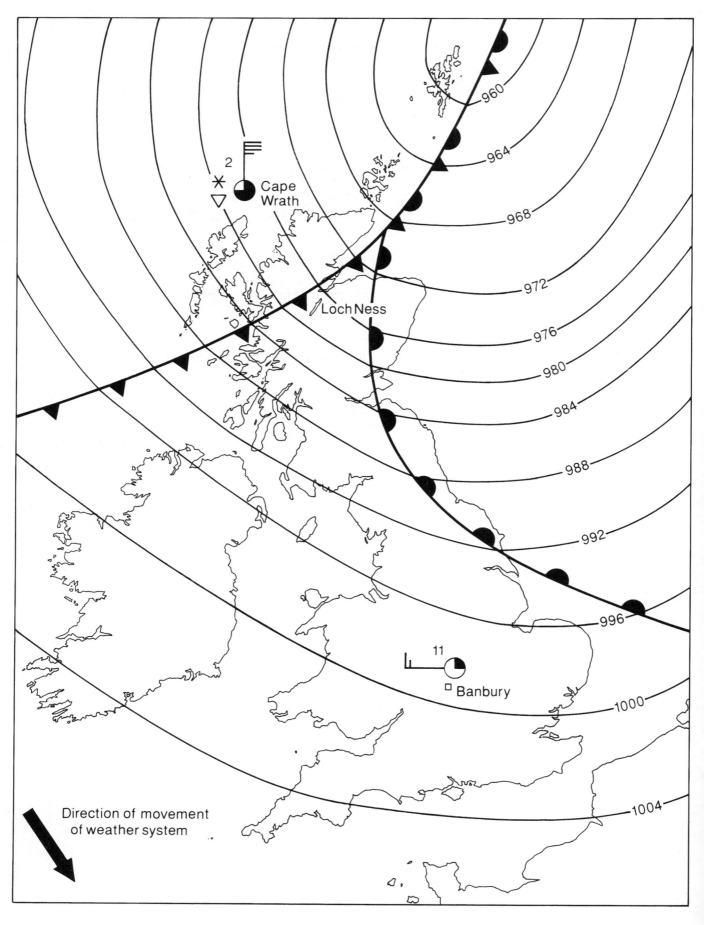

Marks

KU | ES

3. (continued)

Look at Diagram Q3

(*a*) **Describe**, **in detail**, the differences in the weather conditions between Cape Wrath and Banbury at 0800 hours on 10 March.

4

(*b*) At 0800 hours on 10 March a group of secondary school pupils are about to set off on a walk into the mountains near Loch Ness.

Do you think they should go ahead with the walk?

Give reasons for your answer.

4

[Turn over

4.   **Diagram Q4A: Possible Sites for Marine National Parks**

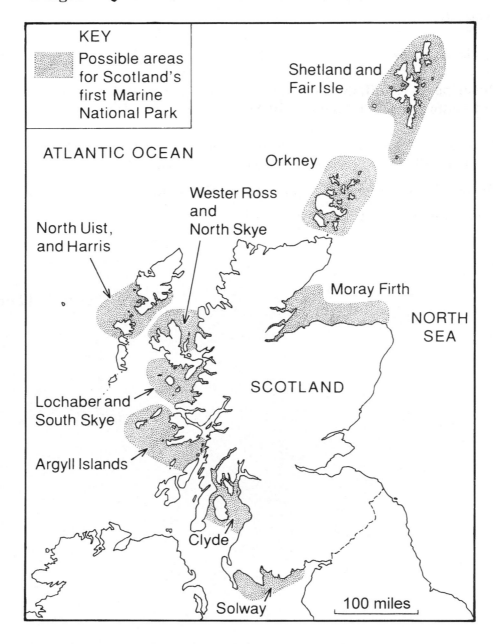

Diagram Q4B: Selected Marine Industries in the Seas around Scotland

 fishing

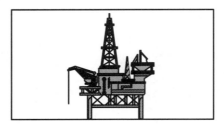 oil drilling

 fish farming

 wave and tidal energy schemes

Marks

KU | ES

4.   **(continued)**

Diagram Q4C: Views about Scotland's First Marine National Park

Oil worker

Environmental protesters

Study Diagrams Q4A, Q4B and Q4C.

There is a plan to establish at least one Marine National Park in the seas around the coast of Scotland.

What are the advantages **and** disadvantages of this plan?

6

[Turn over

Marks

KU | E:

5.   Diagram Q5: Two possible Sites for a Housing Development

Brownfield   **Greenfield**

Study Diagram Q5 above.

On which site would it be better to build a new housing development?

Give reasons for your answer.

Marks

KU | ES

6.

Diagram Q6A: A Lake District Farm

Altitude
250–450 m

Thin soils

Average rainfall
1500 mm per annum

Hours of sunshine
<1000 per annum

Diagram Q6B: Land Use on the Farm

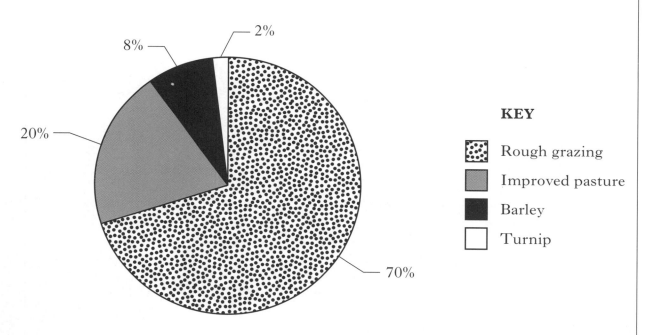

2%

8%

20%

70%

KEY

Rough grazing

Improved pasture

Barley

Turnip

Look at Diagrams Q6A and Q6B.

Explain the land use on the farm.

6

[Turn over

Marks

KU | ES

7. **Diagram Q7: Changing Employment Structures in the UK, 1910–2010**

	Primary %	Secondary %	Tertiary %
1910	15	55	30
1960	5	49	46
2010	2	29	69

Look at Diagram Q7.

What **other** techniques could be used to process this information?

Give reasons for your choices.

5

Marks

KU | ES

8.

Diagram Q8: Extract from Local Newspaper

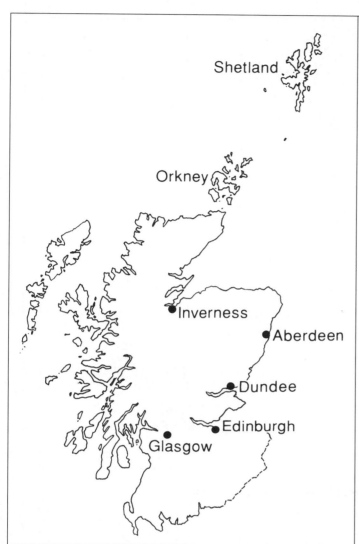

SHETLAND'S CHANGING POPULATION

The average age of those remaining in Shetland will be higher than that of previous years.

*

The number of pensioners is to rise by 52%.

*

The number of children under 15 will decrease by 33%.

*

There will be almost 20% fewer islanders of working age.

Look at Diagram Q8.

Describe the social **and** economic effects these changes will have on remote areas such as Shetland.

5

[Turn over

9.   **Diagram Q9: European Union Member Countries**

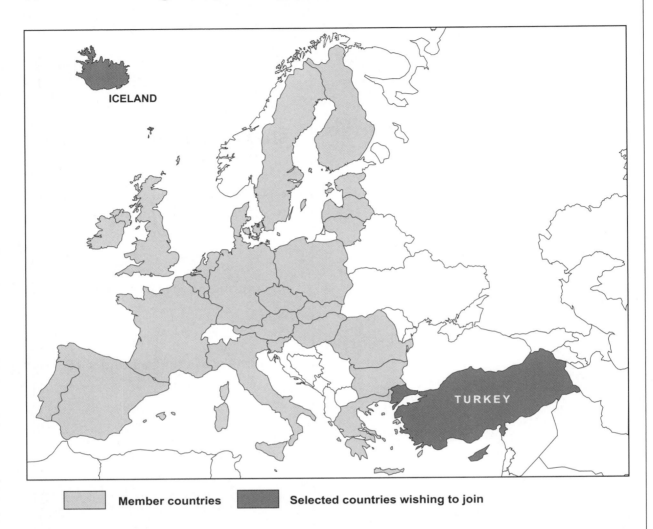

Look at Diagram Q9.

"The European Union continues to grow."

What are the advantages to countries of joining the European Union (EU)?

4

Marks

KU | ES

10.   **Diagram Q10: Fairtrade Products**

Look at Diagram Q10.

Explain the benefits of Fairtrade for Developing Countries.

4

[Turn over for Question 11 on *Page sixteen*

Marks

KU | ES

11.   **Diagram Q11: Different Types of Aid in Developing Countries**

Bilateral Aid	Charity Aid
Tied aid	No conditions
Provides large loans	No loans or debts
Requires repayments	Depends on donations
Leads to increased trade	Provides experienced field workers
Improves links with donor countries	Non-political
Used for agreed projects	Supports small-scale self-help schemes

Look at Diagram Q11.

"Bilateral Aid is better for Developing Countries than aid from charities."

Do you agree fully with this statement?

Give reasons for your answer.

[END OF QUESTION PAPER]

[BLANK PAGE]

Acknowledgements

Permission has been sought from all relevant copyright holders and Bright Red Publishing is grateful for the use of the following:

The European Emblem © The Council of Europe (2009 page 15);

Logo © Action Aid (2010 Credit page 12);

Logo © World Vision (2010 Credit page 12);

Logo © Water Aid (2010 Credit page 12);

The Oxfam Logo, is reproduced with the permission of Oxfam GB, Oxfam House, John Smith Drive, Cowley, Oxford OX4 2JY, UK www.oxfam.org.uk. Oxfam GB does not necessarily endorse any text or activities that accompany the materials. (2010 Credit page 12);

Logo © the Fairtrade Foundation. The Fairtrade Foundation has not supplied the text for the marking instructions (2012 Credit page 15);

Ordnance Survey © Crown Copyright. All rights reserved. Licence number 100049324.

STANDARD GRADE | ANSWER SECTION

SQA STANDARD GRADE GENERAL AND CREDIT GEOGRAPHY 2008–2012

GEOGRAPHY GENERAL 2008

1. (a) *All 4 correct – 3 marks*
   2 or 3 correct 2 marks
   1 correct – 1 mark

Name	Letter of feature	
Angel's Peak	D	Pyramidal peak
Lairig Ghru	C	U shaped valley
Loch Coire an Lochain	B	Corrie
Loch Einich	A	Ribbon lake

3 KU

(b) *Credit should be given for relevant diagrams. Fully annotated diagrams could gain 3 marks.*
Possible answers might include:
Snow collects in hollow then turns to ice (1). This ice eroded the mountain on all sides (1) creating corries (1). These got progressively bigger and their back walls were eroded back towards each other (1). Frost shattering produced a sharp peak (1)
Maximum 1 mark for a list of processes e.g. abrasion, plucking, freezing, thaw (1) 3 KU

(c) *For full marks both points of view should be mentioned. No marks for grid references.*
Mark 3:1, 2:2 or 1:3
Possible answers might include:
Settlement:
Encouraged: flat land along the valley floor making building easy (1), route way through valley (1), beside river for water supply (1), bridging point on river (1).
Limited: surrounded by steep slopes making building difficult (1), possible flooding from the river (1), expansion restricted by the steep slopes (1). Accept reference to forestry (1). 4 ES

(d) *Accept yes/no answers.*
No marks for choice.
Max 1 for list of activities/features.

Possible answers might include:
Good: surrounded by steep slopes for rock climbing and abseiling (1), loch available for water sports e.g. sailing and canoeing (1), forest trails for orienteering (1), access to ski slopes (1), good training area for mountain rescue teams (1).
Bad: very isolated (1), few roads into the area (1), inaccessible at certain times of the year (1), lack of accommodation (1). 4 ES

(e) *Maximum of 1 mark for a list (1).*

Possible answers might include:
Beauty of the countryside destroyed by main road through the area (1), scenery affected by the ski tows (1), beauty taken away by caravan and camp sites (1), litter dropped by tourists (1), views obscured by forestry (1), access restricted by farms, forestry (1), access limited by narrow, country roads (1), local wildlife disturbed by tourists walking on the hills (1), noise from clay pigeon shooting frightens animals

(1), deer killed on roads (1), destruction of local animal habitat by ski slopes (1), rare plants and small animals affected by footpath erosion (1).
Accept conflicts between visitors and farming. 4 ES
Mark 1:3, 2:2 or 3:1

(f) *Techniques might include:*
taking photographs (1), drawing field sketches (1), questionnaire to walkers and cyclists who use the path (1), visit the area and observe (1), interview locals who stay along the route (1), interview park ranger (1). Interview shopkeepers (1).
Maximum of 1 mark for mention of same technique.

Reasons could include:
photographs can be compared if taken at different times to show any changes (1), walkers can give you up to date information (1), and if they use the path frequently can indicate the damage at different times (1), park ranger would have first hand knowledge of impact (1). Shopkeepers can comment on economic impact (1). 4 ES

2. *One mark per valid point, two for developed statements. Credit suitably labelled diagrams.*
Delta: this is formed when the river deposits sediment (1) because the current slows down when it meets the sea and no longer has the power to carry all its load (2). The sediments build up to form islands (1) and the river flows between them in a braided channel (1).
Flood Plain: flood plains are formed when a river bursts its banks and it deposits silt/alluvium on the surrounding land (1). The edges of the flood plain can be made wider due to erosion on the outside bends of the meander (1) and as these meanders move with time, different parts of the flood plain are widened (1).
Credit references to levees and channel deposits.
Ox-Bow Lake: the current of the river erodes the outside bends of a meander (1), this causes the neck of the meander to get narrower (1) until eventually the river breaks through and leaves the old meander abandoned (1). The ends of the meander are sealed off due to deposition (1) leaving a crescent shape or ox-bow lake (1).
There is no credit for reference to meander without appropriate detail. Credit 'deposition on neck of meander' (1) 3 KU

3. *Max 2 marks for correct description of weather.*
No credit for reference to cloud cover or wind direction.

Answers might include:
Snow was forecast (1) which would have made the pitch impossible to play on (1) and would have blocked the roads leading to Pittodrie (1). Temperatures were to be below freezing (1) so the pitch would have been frozen (1) and consequently dangerous for the players (1). Strong winds would have made it difficult to play football (1) and would have caused the snow to drift, blocking access roads (1). 4 KU

4. (a) *2 marks for placing the two lines correctly.*
   1 mark for labelling.

(b) Answers might include:
A: enforce laws banning dumping - this would reduce the amount of leaks from oil tankers (1), force companies to find safer ways of getting rid of waste (1) and perhaps help to encourage more recycling (1).

B: ensure sewage is treated - would benefit marine life (1) and make beaches safer for people (1). It is easier to monitor (1).
Accept negative points about other measure. (e.g. enforcing laws would be difficult (1))   **3 ES**

5. *No marks for naming climatic regions.*
Max 2 marks for description of each climate.

Answers might include:
X: extremely cold temperatures (1) means that most of the year the ground is frozen and cannot be cultivated (1), growing season is too short for farming (1), working outside is difficult in such cold temperature (1) with danger of frostbite (1).

Y: climate is extremely dry (1) making agriculture difficult (1) water is a basic requirement of life so people cannot live here easily (1) very high temperatures can cause heatstroke (1).   **4 ES**

6. *No marks for choice.*
Maximum 1 mark for description or straight lift.

Dairy Farming: close to market for easy transport (1), low cost of transport on a daily basis (1), flood plain good for grazing (1), grass rich on flood plain (1), flat land suitable for dairy cattle as they do not like slopes (1), mild temperatures suitable for cows (1), enough rainfall for good grass (1). Accept negative points.

Arable Farming: adequate rainfall for growing crops (1), good flat land for use of machinery (1), fertile soil saves on using fertilisers (1) close to market for produce (1).   **4 ES**

7. *1 mark per valid point, 2 marks for expanded point. At least two features must be explained.*

Answers might include:
"Close to main road and motorways" means quicker/easier transport of goods (1) which cuts costs for companies (1). It also makes it easier for workers to travel (1).

"Near edge of city" means land is cheaper away from CBD/inner city (1) but site is still close enough to town for a labour supply (1). It has a pleasant environment away from noise/pollution/dereliction in old industrial areas (1)

"Landscaped with grass, trees and shrubs" creates a pleasant environment which helps in the recruitment of skilled workers (1) and also promotes a positive image of the company (1).
"No chimneys" – modern factories tend not to use coal/steam power (1). They use electricity (1).
"Spacious site on flat land": means room for expansion (1) and flat land is easy to build on (1).
"Large areas of tarmac surface" – parking areas for workers who nowadays mostly travel by car (1), parking areas/turning areas/delivery bays since most transport for modern industry is by lorry (1).   **4 KU**

8. *No marks for choice.*
Accept negative points about the alternative.

Route A
Good farmland would not be lost (1). Only a few trees would need to be cut down (1). Does not have to pass through any built up areas (1) so residents would not be worried by noise or accidents (1). There are no other roads nearby so no tunnels or bridges have to bebuilt (1).

Route B
This would be shorter so not as expensive to build (1). Would not need to build cuttings as land is flat (hills along Route A) (1), or cut down trees which destroy wildlife habits (1). Would provide good transport links for farmer taking products to market (1). Could be used by industrial estates to import raw materials and transfer finished products (1) reducing transport costs (1).   **4 ES**

9. (a) *Two factors must be mentioned for 4 marks.*
Accept negative points.
Mark 2:2, 3:1 or 1:3
Answers might include:
Diet: improved diet leads to better health (1) so people live longer (1), balanced diet leads to less dietary diseases (1) e.g. Rickets (1), a balanced diet also reduces the risk of heart disease (1) so increasing life expectancy (1).

Medical Advances: new technology can provide an earlier diagnosis (1), earlier identification allows faster treatment of illnesses (1) reducing the death rate (1), new drugs cure or prevent illness so improve life expectancy (1), advances in treatment of heart disease e.g. heart bypass surgery allow people to survive longer (1).   **4 KU**

(b) *1 mark for relevant technique: Line Graph (1),*
Bar Graph (1)
Any other valid technique with reason.

Reasons could include:
line graph shows trend over time (1) can show rate of change by steepness of the line (1) bar graph shows good visual comparison (1) can be enhanced by colour (1).   **4 ES**

10. (a) *1 mark per valid point, 2 marks for developed point.*
Advantages and disadvantages must be mentioned.
Max 1mark for straight lift.
Mark 3:1, 2:2 or 1:3
Possible answers might include:

Advantages: money can be used to buy seeds or new machinery (1) so that food supply improves (1), improved trade links with EMDCs (1) would allow ELDC to fund projects like HEP schemes (1).

Disadvantages: money may have to be paid back with interest (1) which may result in more debt for ELDC (1), ELDC may not have choice about where to spend money (1), ELDC may have to buy replacement parts for machinery from donor country (1) which means they cannot shop around for the cheapest prices (1).   **4 ES**

(b) *1 mark per valid reason, 2 marks per expanded point.*
Accept yes/no answers.

Possible answers may include:
Yes: better education about family planning would lead to better food supply (1), educating children may help them get better jobs when they grow up (1), being able to read and write means that people can understand leaflets about family planning/
health care (1), earning more money when the children grow up would mean easier living conditions (1) and could help support parents (1).

No: takes some family away from working on the farm (1), needed to grow essential crops/food (1), means traditional way of life will be lost (1), people may be happy with present way of life (1).   **3 ES**

11. *4 marks for six correct, 3 marks for four or five correct, 2 marks for two or three correct, 1 mark for one correct.*

UK
Exports mainly manufactured goods. B
Agriculture employs 2% of population. E
High GNP per capita. F

India
Quotas on exports to Germany and France low. A
Agriculture employs 66% of population. C
Energy consumption per capita is low. D   4 KU

1. (*a*) *1 mark per valid point, 2 marks per expanded point.*
Both river and valley must be referred to for full marks. Max (1) for grid reference.

Possible answers might include:
The River Aire flows in a general NW to SE direction (1) and has a winding course (1) with several meanders (1) eg at 085417 (1); the feature at 063427 may be a cut off (ox-bow lake) (1); the river is joined by several tributaries eg at 043448 (1). Gradient is gentle (1).
The valley floor is wide (1) up to 1km wide in grid square 0841 (1) and flat (1). The sides of the valley are fairly steep (1) and rise from about 90 metres on the valley floor to over 300 metres eg in 0743 (1).
4 KU

(*b*) *A well annotated diagram could earn full marks.*

This is a V-shaped valley eroded by the river in its upper course (1) created by down cutting/vertical erosion (1) by corrosion and hydraulic action (1); the exposed sides are weathered (1) eg by freeze-thaw action (1); particles are moved down the slope by the movement of rainwater/gravity (1) and transported away by the stream (1) eg by saltation and traction suspension (1). 4 KU

(*c*) *1 mark per valid point, 2 per expanded point.*
Answers referring to only advantages or disadvantages can achieve full marks.

Advantages: land to the north of the farm is fairly gently sloping and could be suitable for pasture (1) and even for cultivation using machinery (1). There is access to services in nearby Ilkley (1) and to markets via the A65 primary route (1). Land is well drained because of slope (1).

Disdvantages: there are several footpaths in the area, and the Dales Way nearby, visitors/walkers could be a nuisance (1) location on the outskirts of town could cause vandalism/dumping problems (1) some of the land is very steep, especially to the south of farm buildings making use of machinery difficult (1). Woodland may be habitat for pests eg rabbits (1) land is high, >260 metres so temperatures may be low and it would be hard to grow crops (1). 4 ES

(*d*) *No marks for grid references.*
1 mark per valid point, 2 marks for developed point.

Possible land uses likely to be mentioned would include:

Farming, urban expansion eg south east of Ilkley, local recreation (golf course) and tourist recreation (walking on Dales Way), hotel, parking, sites of historical/cultural interest (eg Pancake Stone).

Possible answers might include:
Conflict between farmers and tourists because farmers complain that tourists damage walls and fences (1), leave gates open allowing animals to stray into danger (1), allow pet dogs to worry sheep (1) and tourists object to farmers' attempts to restrict access (1).
Conflict between urban areas/developers and farmers over spread of urban areas/new housing onto farmland (1) hampering farmers and even putting them out of business (1). Conflict between either of these groups and recreational users such as golfers (1). Conflict between golf course and walkers over access (1).

Conflict between conservationists and walkers over potential damage to historic sites such as Pancake Stone (1). Conflict over traffic congestion between local residents/road users and tourist traffic at busy sites on narrow roads eg near the car park and viewpoint in grid square 1346 (2).
Accept any other relevant points 6 ES

(e) *1 mark per valid point. 2 marks for a developed point. Marks only for differences*

Answers may include:
0440 is an area of newer housing, 0641 an older town centre/inner urban area (1).
0440 mainly suburban residential, 0641 greater variety – town centre, shops, offices and industry as well as housing (2).
0440 varied street pattern including crescents and cul-de-sacs, 0641 less varied (linear/rectangular/grid-iron) (1).
0440 has smaller buildings (houses), 0641 has large buildings including factories/industry (1).
0440 limited amount of traffic, less noise pollution, 0641 many main roads, railway (and station), bus station – more noise and pollution (2).
0440 more open spaces/access to countryside including nature reserve, 0641 less open space and poorer quality environment (2).
Accept other relevant points. 5 ES

(f) *1 mark per valid point, 2 for an expanded statement.*
Good access to network of main roads for cheaper/easier transport of goods (1). This is an important location factor in cutting costs (1); water supply from river (1); flat land for building (1); beside town for labour supply (1).
Area around town has attractive scenery/ recreational opportunities which help to attract workers (1).
Accept other valid points. 5 ES

2. *1 mark per valid point, 2 for a developed statement.*
For full marks two features must be mentioned. Credit suitably labelled diagrams.

Possible answers include:
This landscape was formed by glacial deposition (1), the terminal moraine was formed when a glacier picked up moraine and transported it (1), when the glacier melted the moraine was deposited at the snout of the glacier, marking the furthest point it reached (2).

Beyond the terminal moraine, the outwash plain was formed by melt water streams (1), this transported fine material such as sand and gravel from the ice (1), before sorting and depositing it (1).

Drumlins are small hills made of boulder clay which are moulded by the ice as it passes over (1) the boulder clay is formed beneath the ice as rocks and other debris are ground up and smashed by the erosive effect of the ice sheet/glacier (1). 5 KU

3. *Accept yes/no answers.*

Maximum 1 mark for simple descriptions/list of weather characteristics. No credit for reference to a cold front. If reference to warm front, movement must be mentioned.

Yes: High pressure/anticyclone is covering Britain (1). This could bring settled weather for the whole week (1). Skies will be clear and sunny due to the lack of any fronts (1).

Isobars are well spaced so winds will be light (1). Wind will blow from south bringing warm weather (1), since it is summer temperatures will be hot under the clear skies (1).

No: The centre of the anti-cyclone lies to the east of Britain, suggesting it might move away to the east (1). A warm front is approaching from the west (1). This will bring cloudy conditions and heavy rain due to condensation of moisture in the rising air at the front (2). The isobars will become closer together causing high winds (1).

The high pressure could cause a heat wave (1) with dangers of dehydration for children (1) and sunburn in the sunshine (1). There is a possibility of thunderstorms after a spell of hot weather (1). 5 ES

4. *No marks for straight lifts.*
Accept yes/no answers.

Yes: Carbon dioxide levels in the atmosphere will increase (1) causing global warming (1), raising sea levels by melting of ice caps (1) with world-wide flooding of lowland areas (1).
The valuable hardwood timber is mostly exported, so the world benefits from these resources (1).
Since it has the largest number of plant/animal species of any natural region its destruction will have a major impact on world biodiversity (1).
Since many medicines are derived from plants, losing them will seriously affect research into cures for diseases throughout the world (1).
Cattle ranching and plantation agriculture provide cheap food for export to other countries (1)
Many of the companies exploiting the forest are multinationals, so profits leave the country (1) and Brazil does not benefit much economically (1).
Exploiting the mineral resources, will if they are exported, provide raw materials for many countries, not just Brazil (1).

No: The exploitation of resources like timber and minerals provides jobs for Brazilians and improves the economy of the country (2).
It is only Brazil's wildlife that is affected, not the rest of the world (1).
The HEP provides energy for Brazil's industry (1).
The fact that Indians are forced to live on Reserves only affects people in Brazil (1).
The building of the road means that Brazilian settlers can gain access to virgin forest, providing them with a livelihood (1), and also gives access to resources for Brazilian companies to exploit (1). 6 ES

5. (a) *1 mark for a valid point. 2 marks for a developed point. 1 mark for description. For full marks both zones must be referred to.*

Possible answers might include:
Zone 1 - 19th century
In the inner city tenements/terraced housing were built to save space (1), because this zone is close to the CBD where land is expensive (1) this allowed high population densities (1), houses were close to industry because people had to walk to work (1), little open space or gardens as land was scarce (1).

Zone 3 - late 20th century
Because it is on edge of town where land is cheaper there will be low housing density (1), houses are larger, detached or semi detached with back and front gardens and garages (1), newer housing so better planned layout with cul-de-sacs and crescents (1). 6 KU

(b) *At least two techniques must be described.*
Maximum of three marks if no reasons are given or if reference is made to only one technique. Do not credit the same reason twice.

Possible answers might include:
Comparison of old and new photographs (1) these could be displayed side by side to highlight changes in land use (1) and differences in the amount of open space, building heights and street layouts (1). Photographs could be annotated to show changes (1).

Looking at old and present day maps (1), saves the need for time consuming fieldwork and would show changes in land use (1) and differences in the amount of open space (1), services available then and now (1).

Fieldwork in CBD could record building age, height and function (1), able to record present day land use (1), would be able to compare this with old records, photographs of area (1).

In any of the above give credit for old materials obtained from library/planning offices. **5 ES**

6. (a) *For full marks* **both** *benefits and problems must be mentioned.*

Benefits: Economically such developments create more jobs, so cut unemployment (1) meaning people have a higher standard of living (1), the multiplier effect is likely to kick in (1) so tradesmen and other local firms will get more business from new factories (1), while shops and other services will benefit from local people having more money to spend (1). The economy of the local area will also benefit from increased tax revenue (1), with more money available for spending by the local authority on public services and the environment (1).
Socially, more jobs means people are less likely to leave the area and some more may move in (1), avoiding problems such as depopulation and an ageing population (1). Higher employment levels and higher standards of living are usually associated with a decrease in social problems (1), including crime, vandalism and drink and drug abuse (1) and can reduce stress in family life (1).

Problems: An increase in industrial traffic and movement of workers causing traffic congestion and noise and air pollution (2), firms may want to occupy Greenfield sites with a resulting impact on the environment (1) and if the demand for houses increases, prices may rise rapidly which will cause disadvantage to the less well paid and to first time buyers (2). **6 KU**

(b) *The information in table 1 could be shown by a bar graph (1) or a pie chart (1). Answer must link technique to appropriate table except for bar graph.*

Bar Graph: A bar graph is good at showing individual totals for each category (1). Bars are side-by-side for easy comparison (1) and can be coloured to enhance the presentation (1).

Pie Chart (only appropriate for table 1):
Candidates must refer to figures being converted to percentages before crediting other statements.

Figures could easily be be converted into percentages for each category (1). Pie chart is good at showing proportions (1), colour can be used to highlight the different segments (1).

The information on table 2 could be shown by a line graph (1) or a bar graph (1).

Line Graph: line graphs are good at showing change over time (1) showing not only rising and falling totals clearly but also the rate of change (1) allowing easy comparison between different periods (1).
If bar graph used twice as technique, only one mark. **5 ES**

7. (a) *For full marks at least two stages must be described.*

Possible answers include:
Birth Rates: In stage 1 these are high (over 40/1000) (1) remaining high in stage 2 (1) they fall sharply in stage 3 to around 10 per 1000 (1) levelling out and remaining low in stage 4 with one or two small baby booms (1).

Death Rates: These were high and fluctuating in stage 1, between 43/1000 to 46/1000 (1) falling sharply in stage 2 to around 11/1000 (1) remaining low in stages 3 and 4 (1).

Total Population: This remained low in stage 1 and into stage 2 (1) halfway through stage 2 total population started to increase (1) total population increased dramatically in stage 3 (1) levelling out in stage 4 (1). **4 ES**

(b) Possible answers include:
Nigeria: In Nigeria there is little contraception available so birth rates are high (1), high numbers of births to ensure some children survive (1), children are needed for an income (1) and to look after parents in their old age (1) as no government pensions (1).

Death rates are falling due to Primary Health Care (1) foreign aid (1) availability of medicines from developed countries (1) high birth rates and falling death rates means a large natural increase (1).
Any other valid point.

UK: Low birth rates and low death rates mean very low natural increase (1). Low birth rates results from better medical care of mothers and babies (1), so most babies survive (1) so lower birth rate (1), low birth rate from use of contraception (1), birth rates low from later marriages (1), children are expensive (1), women want careers rather than children (1). Low death rates due to medical advances (1), good living conditions (1), National Health Service (1).
Accept other valid points. **4 KU**

8. (a) *1 mark for straight description of graphs.*
1 mark per valid difference, 2 marks per expanded point.

Possible answers might include:
Exports mainly manufactured goods – 97% but imports quite a lot of raw materials (1) like food and oil (1), doesn't seem to export any raw materials (1). 38% of imports are raw materials (1) only possible 3% of exports are raw materials (1). **3 ES**

(b) *1 mark per valid point, 2 marks per expanded point.*

Possible answers might include:
Japan has very little in the way of natural resources of her own (1) so has to import them (1) eg oil and timber (1), relies on making goods using bought resources (1) to be able to make money (1), will not pay highly for raw materials but can sell electrical goods for high prices (1) so that businesses make large profits (1). **3 KU**

GEOGRAPHY GENERAL 2009

1. (a) *For full marks, reference must be made to river **and** the valley. Mark 3:1, 2:2 or 1:3. One mark for correct grid reference.*
The River Sowe flows mainly south west (1); it has several large meanders (1); the floodplain is quite wide (1), and there is a lake on the floodplain (1) in 3778 (1). River is quite wide (1).
There is an island in the river at 340753 (1). There is an ox bow in the making (1).
There are steep slopes on the eastern bank in 3474 (1).
Or any other relevant point. 4 KU

(b) Mark 3:1, 2:2 or 1:3. No marks for grid references.
Advantages: There is attractive scenery including a lake and forestry (1) which means people can do a number of different activities (1) such as water sports or orienteering and walks in the forest (2). It is close to Coventry so available to lots of people (1) and it is easy to get to with an A and a B class road nearby (1). There are sites of interest to visit such as Combe Abbey (1).
Disadvantages: A dual carriageway runs through area which might cause noise and air pollution (1); some parts of area A are built up which would spoil the scenery (1) and would not be so good for wildlife (1). There are several farms in the area which might create conflict with visitors to the Country Park (1).
Or any other valid point. 4 ES

(c) One mark per valid statement or two for a developed point.
For example:
Many main roads converge on this square (1) and there is a tourist information centre (1). A Cathedral (1) and a bus station (1) are also located in 3379. There are several churches here which often indicate the old centre of a city (1). 4 ES

(d) *One mark per valid statement or two for a developed point. Accept negative points about the other grid square. No marks for grid reference. Maximum 1 for list.*

Area X <u>3278</u>: this area is conveniently close to the CBD for work and shops (1) so travelling costs would be less (1).
This area is in the inner city of Coventry so there may be small terraced houses which are quite cheap (1); there is a school for the couple's children in 3278 (1).

Area Y <u>2778</u>: this area is on the edge of Coventry and would have a nicer environment than 3278 because there is more open space (1); the houses are likely to be more modern, have bigger gardens and be suitable for families (2); there are woods to walk in (1); there are fewer main roads so there will be less traffic than in 3278 (1); the couple could commute to work from the railway station just south of the square (1). 4 ES

(e) *One mark per valid statement or two for a developed point. Mark 1:3, 2:2 or 3:1 Maximum of 1 for grid reference.* For example:

Advantages: the land here is quite flat and so easy for machinery to work on (1) it will be easy to get workers being so close to Coventry (1) and there will be a big market on the doorstep so the farm could produce milk or fruit and vegetables which will not have to be transported far (2) it is close to the M6 for easy transport of farm produce to other cities (1).

Disadvantages: It is too close to the M6 which will cause a lot of noise and air pollution (1); there are electricity pylons running across the farm which will be ugly (1); it is too close to housing estates so there could be problems with vandalism (1); the farm will be under pressure from urban expansion (1); the access to the farm is quite poor as there is only a small track nearly a kilometre long (1). The pylons make it difficult to move machinery (1). 4 ES

(f) *One mark per valid statement or two for a developed point. Allow yes/no answers.* For example:

Yes: there will have been more jobs provided in the area, both during the construction and once the stadium is running (2); demolishing the gasworks has got rid of an eyesore (1); the stadium is right next to a dual carriageway and close to the motorway so fans will be able to get there easily (1) without having to go near the city centre so reducing traffic there (1); there is space around it for car parking (1).

No: the stadium is too far away from the station so fans will have to travel by road (1) increasing the amount of traffic in the area (1). It is too close to housing areas so people in this area will now be disturbed by noisy football fans (1). It is not very central and only convenient for people in the north of Coventry (1). 3 ES

2. *Fully annotated diagrams can obtain full marks.*

Ox bow lake: As meander gets bigger (1) the neck narrows (1) due to erosion on the outside of the bends (1); material is deposited on the inner bends (1). Eventually river becomes straight with meander cut off to form ox bow lake (1).

Waterfall: Soft rock overlain by harder rock (1). As water flows down valley it erodes the softer rock faster than harder rock above (1). Eventually soft rock cannot support harder rock (1) which collapses forming a waterfall (1). 4 KU

3. *One mark per valid statement or two for a developed point. One mark for correct identification of weather station Y. Credit statements explaining why it cannot be X. For full marks reference must be made to synoptic chart.*

Symbol Y (1)

eg Bordeaux is in the centre of a high pressure system (1) so will have little cloud cover and no rain as there are no fronts in an anticyclone (2); isobars are widely spaced so there is little or no wind at Bordeaux (1); temperatures are quite high as it is noon at the end of March (1); it will be sunny as there is only one okta of cloud cover (1). Anticyclone usually has clear skies (1).

4 KU

4. *(a) 1 mark for each bar correctly plotted on the graph.* 2 KU

(b) Answers must refer to both temperature and rainfall for full marks.

eg Temperatures vary from 12 to 25 degrees centigrade (1); the range of temperatures is 13 degrees (1); the highest temperatures are in June and July – 25 degrees (1) and the lowest temperature is 12 degrees in December (1). Rainfall varies throughout the year (winter maximum); the highest rainfall is 80mms in November (1) and the lowest rainfall is 20mms in July (1). Winters are cooler and wetter than summers (1); this graph shows a Mediterranean type climate (1). Total rainfall 618 mm. 4 ES

5. *Maximum of 1 mark for list.*

Unpredictable rainfall or no rainfall (1); high evaporation rates (1); wind blowing topsoil away (1), heavy rain (flash floods) washes soil away (1). Overgrazing by cattle/sheep/goats (1), overcultivation due to population pressure (1) or soils lacking nutrients (1); clearing of woodland for firewood/building materials (1); soil has no vegetation to bind it together or protect it from elements (1).

Or any other relevant point. 3 KU

6. *(a) One mark per valid statement or two for a developed point. Answers should be explanations.*

For example:
This is a large site, so plenty of space for companies and for expansion (1). Good road transport such as the M6, allowing quick and easy access for components and workers (2) excellent communications via the M6/M1 to the North and London to the South (1); centrally located in the UK for the distribution of finished products so reducing transport costs (1). Close to Birmingham airport allowing management easy access (1). There is a large population living within 30 km drive so access to a large pool of workers (1) may be close to colleges and universities so graduates available for management positions (1).

Accept any other valid point. 4 ES

(b) Mark 2:2
Award up to 2 marks for each reason.

Technique	Justification
Questionnaire to local residents (1)	Can find out if the Park created jobs in the area (1).
Interview owners/managers of businesses in the Park (1)	They can tell you where their workers come from (1).
Contact the local council (1)	They would be able to give employment figures for the Park (1).
Traffic survey (1)	Show the impact of traffic on the local area (1); show the amount of traffic using the park (1), increasing congestion (1) and pollution in the local area (1).
Photographs (1)	Show visual impact (1) Show congestion (1)

4 ES

7. *No mark for choice of measures.*
Mark 1:3, 2:2 or 3:1.
Only credit the same reason once.
Credit negative points about measure not selected.

Ring Road:	Allows through traffic to avoid the city centre (1), thus reducing the amount of traffic in the centre (1).
One-way street system:	Prevents jams caused by cars trying to turn right across the flow of oncoming traffic (1), so it allows traffic to flow more smoothly (1).
Parking restrictions:	Deters drivers from bringing their cars into the city (1), reducing traffic (1). Yellow lines make it illegal to park (1), so there is more room in the road for moving traffic (1).
Flexitime working:	People will not all have to travel to work at the same time (1), reducing the number of cars on the road in the rush hour (1).
Park and ride schemes:	Encourages drivers not to bring their cars into the centre (1), reducing the number of cars (1).
Multi storey car parks:	Enables cars to be parked off the road (1) so that traffic can flow freely (1).

Accept any other valid point. 4 ES

8. *One mark per valid statement or two for a developed point.*
Maximum of 1 mark for straight lift.
For example:

Advantages:	areas of the UK have had developments funded by regional aid (1); free trade means a bigger market for UK companies (1); UK citizens free to work anywhere in Europe – more opportunities (1); protection from competition from outside Europe (1).
Disadvantages:	UK contributes more than it gains (1); competition from European companies in UK markets (1); UK cut off from traditional trading partners (1); unrest caused by arrival of "foreign" workers (1). Problems with the value of the Euro (1).

4 KU

9. *(a) One mark per valid statement or two for a developed point.*

The number of people in the youngest age groups will decrease greatly (1), the birth rate will drop (1). The number of people in the oldest age groups will increase greatly (1), life expectancy will increase (1). The number of people in the in-between (working) age groups will decrease, with the biggest drop in the 30-34 group (2). The dependent population will be increasing (1). Death rate will drop (1).

Accept any other valid points or accurate detail of change. 3 ES

(b) One mark per valid statement or two for a developed point.

For example:
There will be fewer children so some schools may have to close down (1), so teachers will lose their jobs (1); the

government won't have to spend so much on education but it will need more money to look after the extra old people (2); this will have to be spent on pensions and building more old people's homes (2). There will be fewer people of working age so there might be a labour shortage (1).   **4 KU**

10. *At least **two** techniques must be given for full marks, so mark 2:2 or 3:1.*

Techniques	Justification(s)
Pie chart(s) (1)	The figures are already in percentages (1) and two pie charts side by side would allow easy comparison (1)
Multiple bar graph (1)	Different colours could be used for imports and exports so you could see the differences – they will be right next to each other (1).
Pictograms (1)	A pictogram for imports and one for exports would be needed (1) the pictures on the pictogram might be easier to interpret without having to look at a key (1).
Classify as a table (1)	Figures for imports and exports would be next to each other, and so easy to compare (1); they could be placed in rank order so you could easily see which was the biggest (1).

     4 ES

11. *One mark per valid statement or two for a developed point.*

Use of local materials (1) means no need for expensive imports (1), so not only richer farmers can afford it (1); more efficient farming creates a better food supply (1), and may leave a surplus to sell (1), improving living standards (1); simple/appropriate technology (1) reduces dependence on outsiders (1); freedom from loans/debt and interest payments (1) means money can be spent on development (1). Stone lines prevent soil erosion (1) hand pumps help to improve water supply (1).   **3 KU**

GEOGRAPHY CREDIT 2009

1. (*a*) (i)

pyramidal peak	711130
corrie	715123
truncated spur	733110
hanging valley	723125

4 correct = 3
2/3 correct = 2
1 correct = 1   **3 KU**

(ii) *One mark per valid point.*
Fully annotated diagrams may gain full marks.
Credit references to frost shattering, abrasion, ice plucking and interlocking spurs (latter only relevant to truncated spur).

Truncated spur – is formed when the slope of a hill is eroded by a glacier (1) as the ice moves down the valley it abrades the sides of hills (1) and ice at the edge of the glacier freezes onto the rock and plucks it away (1). When the ice melts, the slope is left as the steep side of a u-shaped valley (1) and may have crags or cliffs where erosion was greatest (1).   **4 KU**

(*b*) *For full marks at least one similarity and one difference should be mentioned – if not mark out of 5.*
Max 2 marks for evidence relating to any single function.
Differences which are implied should be credited.
Max 1 for grid reference.

Similarities
Both are holiday resorts (1) because both have a tourist information centre and a leisure centre (1), a cycle route passing through (1) and a caravan/camp site (1). Barmouth has a sandy beach which would attract holiday makers and tourist facilities such as a museum while the attractions of Dolgellau are the surrounding hills and forests for walking (1). Both have footpaths leading out of town onto the surrounding hills (1).

Both towns are service centres (1) because both of them have hotels, leisure centres and churches (2).

Differences
Dolgellau is a route centre (1) with 5 main roads meeting there (1) Barmouth has only the A496 passing through (1). Barmouth is a port (1) with a harbour and a ferry (2).
Accept any other valid point.   **6 ES**

(*c*) *One mark per valid point*
Answers should be explanation

A minor road runs directly into the heart of the area (1) allowing heavy logging machinery and workers access to the plantations (1). It connects to an A class road, allowing the felled trees to be transported to market (1). There is a workforce available in Barmouth, about 4km away (1). The slopes are mainly south facing, giving warmth, allowing trees on higher ground to grow better (1) and sloping land is well drained (1). The height of the land makes it too cold for farming but trees can grow in the cooler conditions (1) trees can survive on the moderate slopes where the soil is likely to be poorer and where little else can grow (1) the slopes are not suitable for arable farming as they restrict the use of machinery (1).

Accept any other relevant point   **4 KU**

(d) *Mark 3:1, 2:2, 1:3*
Max 1 for grid references.

Advantages
It passes through fine scenery (1) with good views of mountains at first, and later of the sea (1). There are varied habitats – moors, woodland, river and sea for wildlife watching (2). There is a picnic site (1) at 697153 (1) where they can stop for a snack. There is a telephone box at 698152 from which they could call an ambulance if there is an accident (2). It does not pass through any settlements, so will be peaceful (1).

Disadvantages
It climbs up to 240m (1) so will be strenuous (1). At this height it might be quite cold (1). The descent from 658145 is very steep, so could be dangerous (2). There are very few places on the route where they could buy provisions (1). 4 ES

Accept other relevant points.

(e) *At least two techniques must be described. Maximum of three marks if no reasons given, or if reference is made to only one technique.*
Mark 3:2, 2:3

Visit site and take notes/annotate map (1)	Allows the features of both areas to be noted and compared in greater detail (1) and provides up to date first hand information (1).
Field Sketch (1)	Can be selective on information gathered (1).
Take photographs of both areas (1)	If displayed side by side similarities and differences easily noted (1) they can be annotated to emphasise particular features (1) they can show the areas in greater detail (1).
Extract information from an OS map (1) (Specify type of map)	This would show up differences in the size of the area or land use (1) since the two areas are a distance apart this would allow the areas to be compared without the need for travel cost or time (1).
Write to the National Trust (1)	Since the National Trust own/look after both areas they will be able to provide facts and figures on both areas allowing accurate comparisons to be made (1).

Or any other valid techniques and justifications.   5 ES

2. *One mark per valid point, 2 for a developed point.*
Credit relevant diagrams. Credit reference to process

In the upper course of a river, water flows quickly through a narrow channel, with a steep gradient (1). It uses this energy to deepen its bed (1) this is called vertical erosion (1). The river carries stones and rocks in the water, and the force of the water and the grinding of rocks and stones by abrasion cut down into the river bed (2). Rocks on the valley sides can be broken down by freeze thaw or chemical weathering (1) and mass movements carry this loose material down the valley and into the river (1) the river transports this material downstream (1).

Accept any other valid point   4 KU

3. *One mark for a simple point, two marks for a developed point.*
Mark out of 5 if candidate has misidentified fronts but explained weather correctly.

Answers could include:

As the warm front approaches Glasgow air pressure will fall (1) cloud cover will increase (1) and steady rain will fall (1) winds will be quite strong as the isobars are close together (1). The warm front will move away and Glasgow will be in the warm sector of a low pressure system (1). Temperatures will rise and it will be mild with occasional showers and some cloud cover (1). Winds will die down (1). The cold front will arrive and cloud cover will increase with cumulonimbus clouds bringing heavy rain to the city (2). Temperatures will drop as the cold front passes over and begins to move away (1) the sky will become clear (1) the rain will stop (1) and pressure will begin to rise (1) and winds will increase (1).

Accept any other valid point.   6 KU

4. *Accept yes/no answers.*
Max 1 marks for straight lifts

Yes
Jobs created reducing unemployment (1) boosts the whole of the Scottish economy (1) reduction of pollutants from coal fired power stations (1) increased use of renewable energy (1) conserving the earth's scarce resources (1) pylons built mainly in remote areas so fewer people will see them (1) much less expensive than underground cables (1).

No
Natural beauty of the landscape is scarred by pylons (1) pylons are unsightly and conflict with the aims of National Parks (1) although pylons are already present in the Cairngorm National Park the proposed pylons are more than twice the size of the normal pylons (1) so causing far greater visual pollution (1) disturbs ancient burial grounds (1) very expensive to build and maintain (1) underground cables are more expensive but would not create visual pollution (1).

Or any other valid point.   6 ES

5. *One mark for a simple point, two marks for a developed point.*
No marks for description.
Mark 4:2, 3:3, 2:4
Possible answers include:

There are more shops and offices in the CBD due to its greater accessibility (1) as a result of there being stations located there (1) and A class roads converging (1). Land values are high in the CBD so only large shops and offices can afford the rents (1). There are hotels in the centre because they are close to the stations (1). There is much less open space in the CBD due to the high demand for land (1). The residential land is mainly in the suburbs because land is cheaper there (1) and the environment is more suitable with cleaner air and less noise (1). There is industry in the suburbs because new industries are located at the edge of cities for ease of access by road, avoiding the traffic congestion of the centre (2).   6 KU

6. *No marks for choice. Accept yes/no answers.*
1 mark per valid point. 2 marks for a developed point.

Agree
Machines can do jobs faster and more efficiently than farm workers (1) Larger fields have made it easier to use machines and help increase yields (2) increased yields increase the farmer's profit (1) and makes the country more self sufficient (1). Fewer farm workers saves the farmer paying wages (1) and he can use the empty farm workers cottages to get extra income by renting them out to tourists (1) Surplus food production has been reduced by using set aside (1) and the farmer is paid by the government/EU for doing this (1).

Disagree
Machines are large and spoil the natural look of the countryside (1) and also pollute the atmosphere (1). Larger fields have meant hedgerows have been removed destroying the natural habitat of birds and animals (1) disease is more easily spread in large fields (1). Farm workers have been replaced by machines

so farming jobs have been lost (1). Increased yields can lead to a food surplus (1) and to reduce this, land is left uncultivated (1) or farmers are given quotas to reduce output (1).   5 ES

7. *Three factors need to be mentioned for full marks.*
   If not mark out of four.
   Example: Research Links
   Industries on the science park can make use of research carried out at nearby Southampton university (1) and use the research facilities of the university to develop and improve their products (1).

   Example: Skilled Workers
   There will be a highly qualified supply of graduates from the university available for employment (1) opportunities for student placements to enhance and bring new ideas to their business (1). Skilled workers will be available from the many towns nearby eg Winchester, Basingstoke (1).

   Example: Good Access
   Good road communications to allow workers easy access to and from their work (1) close to M27 and M3 allowing parts and products to be easily transported and distributed (2) a rail service to Waterloo allows staff access to the facilities of London, avoiding the traffic problems of the city (1). Close to Southampton and Heathrow airports allowing staff access to conferences and meetings in the UK, Europe and World destinations (1) and worldwide transportation of products (1). Close to the ports of Portsmouth and Southampton allowing components to be shipped in and out by sea (1).   5 ES

8. *One mark for each relevant point. Two marks for developed statements.*

   eg Population is not evenly distributed across Europe (1) some areas such as England are quite densely populated (1) while other areas, such as most of Scandinavia, have low population densities of under 50 people per square kilometre (2). Some parts of the coastline seem to be quite densely populated (1) such as Spain and Portugal (1). Areas with mountains have quite low population densities such as Scotland, the Alps and the Pyrenees (2).   4 ES

9. (a) *One mark for each relevant point. Two marks for developed statements.*
   For full marks reference must be made to both graphs.

   eg The percentage of the population living in urban areas is expected to increase in both ELDCs and in EMDCs (1). The percentage will rise to 83% in EMDCs by 2030, whereas in ELDCs it will increase to 56% (1) there is a bigger proportional increase in the ELDCs (1).

   Overall the total world population is rising but it is going up faster in urban than in rural areas (2). Between 1950 and 2030 the world's urban population will have gone up about 5 times while the rural population will have increased by less than 50% (2) this shows a very dramatic rise in urban population (1).   4 ES

   (b) *One mark for each relevant point.*
   Two marks for developed statements.
   Marks can be given for explanations about why birth rates are high and for explanations about rural-urban migration.

   eg Urban populations will continue to increase because there is a high birth rate in many parts of the world (1) and also because there will be rural-urban migration (1). Couples continue to have large numbers of children because there is poor access to information about family planning in some ELDCs (1) and also because they may want to ensure that at least some of their children survive into adulthood and are able to look after their parents when they are elderly (2).

Many families in ELDCs continue to move from the countryside to the cities because of the problems in rural areas such as famine, lack of health care and poor job opportunities (2). They move to the cities because they may be able to earn higher wages, get better jobs and ensure their children have a proper education (2). In EMDCs most migrants move to the cities (1).   5 KU

10. *Give marks for different techniques and for justifications of the chosen techniques. Mark 2:3 or 3:2.*
    Don't credit same justification twice.

    Possible techniques include;
    Pictogram (1) pie charts (1) bar graphs (1) divided bar graphs (1) line graph.

    Justifications
    Pictogram: proportional drawings of old people could be used to show the different percentages in Japan and Nigeria (1) the larger drawings for Japan would give/allow clear visual comparison between the two countries (1).

    Must be plural   Pie charts: a series of pie charts would show the changing proportions of old people in Japan and Nigeria (1) the data would suit pie charts well because it is already in percentages (1) and the section of each pie chart showing population over 65 could be highlighted to show the differences between the two countries (1).

    Bar graphs: two bar graphs could be drawn to show changes in the over 65 population firstly in Nigeria and secondly in Japan (1) alternatively they could be shown on the same graph with different coloured bars for each country to emphasise the changes from year to year (1).

    Must be plural   Divided bar graph: a series of individual divided bar graphs for each country would show the changing proportion of old people over time (1) the section for the over 65's could be highlighted to allow clear comparison (1).

    Line graph: the changes over time could be shown easily on a line graph (1) different coloured lines for each country would show the different trends (1).

    Or any other relevant technique.   5 ES

11. *Accept answers which agree, or which disagree, or which consider both sides. No marks for straight lifts.*

    One mark for a relevant point, two for a developed point.

    eg if "Agree"
    no loans/debts means no interest to be paid (1) and the country's resources can be developed for its own benefit (1) no conditions means people can shop around for the best deal on materials (1) experienced field workers have knowledge of successful schemes elsewhere, and are committed to promoting self-sufficiency (2).

    Accept negative points about Bilateral Aid.

    eg if "Disagree"
    Depends on donations, so regularity cannot be guaranteed (1) high profile events/disasters may cause funds to be diverted (1) amount of money available is less than from governments, and cannot fund even appropriate large-scale developments (2).

    Accept positive points about Bilateral Aid.   4 ES

GEOGRAPHY GENERAL 2010

1. (*a*) *3 marks for 4 correct; 2 marks for 3 or 2 correct; 1 mark for 1 correct.*

A592	D
Brown Cove	A
Mixed Wood	C
Red Tarn Beck	B

3 KU

(*b*) (i) *3 marks for 4 correct; 2 marks for 3 or 2 correct; 1 mark for 1 correct.*

Glacial Feature	Grid Reference
Hanging valley	3108
Corrie with Tarn	2807
Misfit Stream	3006
Ribbon Lake	3115

3 KU

(ii) *1 mark per valid point. 2 marks for a developed point. Well annotated diagrams could obtain full marks. 1 mark for list of terms.*

Eg a corrie

Snow gathers in a north-facing hollow on the side of a mountain (1).

Snow accumulates and becomes compressed into ice (1). Ice moves downhill under gravity (1). Plucking and abrasion steepen and deepen the hollow (2). When the ice melts a deep armchair shaped hollow called a corrie is left (1).

Or any other valid point.   **3 KU**

(*c*) *No marks for choice. 1 mark for a simple point, 2 marks for a developed point. Yes/no answers acceptable. For full marks at least one piece of map evidence must be given.*

If **YES** chosen

No marks for grid references.
There is flat land on the valley floor for building on (1). It is well drained so little risk of flooding (1). There is access via the A593 and B5343 (1). There are some services nearby like the hotel at 285061 (1). The glaciated scenery offers a variety of activities for visitors to the area (1) eg walking, cycling, sightseeing (1). Create jobs in the area (1). Reference to appropriateness of building design (1).

If **NO** chosen

Steep land is hard to build on (1).
The area is part of the National Park so scenery would be spoiled (1).
There is only one B class road into the area (1). Wildlife habitats may be destroyed by lodges (1). Lodges would encourage more visitors into the area who could drop litter and erode footpaths (2). Area lacks services (1).

Or any other valid point.   **4 ES**

(*d*) *Yes/no answers acceptable. 1 mark for each relevant point. 2 marks for a developed point. No marks for grid references.*

Yes: Flat land to south and north west (1). Flat land is easy to build on (1) and is at the bottom of a valley so is sheltered (1).

No: Much of the land from north to south east is high and steep so the settlement's expansion here is restricted (1) as building here will be difficult/expensive (1). Houses on the high/steep land could be at the mercy of strong and cold winds (1). The low flat areas could suffer from flooding (1).

Or any other relevant point.   **4 ES**

(*e*) *Yes/no answers acceptable. No marks for grid references.*

Agree

Gently sloping land so some crops can be grown (1). Higher land can be used for sheep (1). There is town nearby for a market (1).

Disagree

Rough grazing will be found and sheep can survive on this (1); the flat land cannot be used for crops because it floods (1). Land is very steep so crops cannot be grown (1) or machinery used (1). The soils will be thin and infertile so not rich enough for dairy farming (1).

Accept any other valid point.   **3 ES**

(*f*) *1 mark for each relevant point. 2 marks for a developed point. Maximum 1 mark for Grid Reference.*

Answers might include:

Areas of woodland at 4013 (1) and 3816 will have to be removed (1) which will be bad for wildlife (1). The youth hostel and hotel in grid square 3915 will have to be demolished (1) destroying families' sources of income (1). Disturbance for the settlements of Bridgend at 3914 and Patterdale at 3915 is likely to occur (1). It will be very expensive (1) because it will be about 12.1 km long (1). Grisedale Bridge at 3916 may need to be made wider (1). Brothers Water in 4012 will make expansion west difficult/add to the expenses (1). The valley is very steep and narrow in some places therefore not much room to expand the road (1). Cuttings will have to be made in some area (1) eg at grid reference 4009 (1) the sides of which have to be made safe from falling scree (1). Low lying land will be prone to flooding (1). Farms such as Noran Bank Farm in grid square 3915 may disappear (1), Kirkstione Pass Path will have to be moved (1). This development will scar the beautiful landscape (1).

Accept any other relevant point.   **4 ES**

2. (*a*) *3 marks for 4 correct; 2 marks for 2 or 3 correct; 1 mark for 1 correct.*

3 KU

(*b*)

Location	Weather System
British Isles	Depression
Spain	Anticyclone

1 mark for correct completion of table.
British Isles has fronts (1); isobar values over Britain lower than 1000mb (1); getting rain (1); also windy (1); whereas isobars over Spain over 1000mb (1); and far apart (1); Spain also has obscured sky, fog being common in anticyclones (1).   **4 KU**

3. (*a*) *4 correct, 2 marks; 2 or 3 correct, 1 mark; 1 correct, no mark.*

	Map Area	Graph
Hot Desert	C	
Equatorial Rainforest		4
Mediteranean	B	
Tundra		3

2 KU

(*b*) *For full marks both temperature and precipitation must be mentioned.*

Answer might include:

There is high rainfall throughout the year (1); over 150mm every month (1). The highest rainfall is around 250 mm in January (1). The lowest is approximately 150 mm in July (1).

Temperature is high throughout the year (1). The maximum is 27°C in May (1). The minimum is 25°C in January/November/December (1). The range of temperature is 2°C (1).   **3 ES**

4. (a) *For full marks both advantages and disadvantages must be mentioned. Mark 3:1; 2:2 or 1:3.*

If Retail Park is chosen

Advantages

Near roads for good access (1) near housing for customers (1) near housing for labour (1) space for expansion (1) land possibly cheap (1). Bring jobs to the area (1).

Disadvantage

Building on greenbelt may lead to objections/ conflict and possible hostility (1) and may cause restrictions on further expansion (1); supermarkets or a retail park may suffer from competition from stores/services in Glasgow (CBD or elsewhere) (1); travel between built up areas and the retail park may be difficult because of heavy traffic/congestion (1) especially as there are so many major junctions/intersections nearby (1).

Accept any other valid point.   **4 ES**

(b) *1 mark for each valid technique, 2 for reasons. Both "reasons" marks can be gained from the same technique.*

Possible techniques and reasons:

Technique: Take photographs (1) and compare with old photos obtained from library (1).
Reasons: Putting photos side by side will show differences clearly (1). Photos can be annotated to highlight differences (1).
Technique: Issue questionnaires to/interview people who have lived/worked in the area for a long time (1).
Reasons: Those who know the area well can provide accurate information (1) and can give opinions on change as well as factual detail (1).
Technique: Draw a land use map and compare with an old land use map/photograph from library (1). Draw a field sketch (1) and compare with recollection of older residents (1).
Reasons: This will show location and extent of changes accurately (1). Colour can be used to highlight changes (1). Areas of conflict can be identified (1).

Accept any other valid technique/reasons.   **4 ES**

5. *1 mark per valid point, two for a developed statement.*

Answers may include:

Local people may benefit from the many jobs (1) created in building the new developments (1); after the games are over, the people of East London will have excellent sports and leisure facilities (1) and the Olympic Village could be turned into new housing for local people (1); the new railway station will improve communications in the area (1); the area will get massive publicity from staging the Olympics and many visitors to the areas will give local hotels, restaurants and other businesses a huge economic boost (2).

Or any other valid point.   **4 ES**

6. *1 mark per valid point, 2 for a developed statement. No marks for description. Yes/no answers acceptable.*

Answers may include:

Yes: The farmer has tried to diversify (1); this will bring more money into the farm and make the farm less dependent on producing food (2); the renovated cottages look better (1); the set-aside land and the farm woodland will encourage wildlife (1) and this is good for the environment (1); tourists will be able to see wildlife on the nature trail (1); there are plenty of things for the tourists to do such as mini-golf and quad bikes and this will encourage them to visit the farm (2); they will spend money in the tea room/farm shop (1); all this will benefit the farmer (1).

No: The changes have taken away a lot of useful land (1) and so the farmer has had to cut back the number of sheep and beef cattle (1); the farm now depends too much on tourists (1) and there may not be that many as it is only on a single track road and 120 kilometres to the nearest town (2), making access difficult (1); some of the changes are not good for the environment such as the new car park and the quad bike circuit which will create noise and air pollution (2); there will be hardly any tourists in winter, so the farm will take in very little money then (1).

Accept any other valid point.   **4 ES**

7. (a) *4 correct; 3 marks. 2 or 3 correct; 2 marks. 1 correct; 1 mark. Age groups must be drawn reasonably accurately.*   **3 KU**

(b) *1 mark for a simple point. 2 marks for a developed point. Comparative points should be made.*

Angola is an ELDC and Belgium is an EMDC (1) so Angola does not have the money to improve the standard of living and life expectancy of its people (1). A high % of people are employed in agriculture in Angola compared to the people of Belgium who are mainly employed in secondary/tertiary industries. These generate more money for the people of Belgium (2). In Angola there is a lack of clean water so many people die from water related diseases whereas people in Belgium have access to clean piped water (2). Fewer people live in towns and cities in Angola so do not have easy access to schools, doctors, and hospitals compared to Belgium (2) so many people die through lack of knowledge and medical care (1). It has very few products to trade with and generate money to improve the country's standard of living (1) compared to Belgium which has many products so can trade generating wealth (1).

Accept any other relevant point.   **4 KU**

8. Technique: Bar Graphs (1)
Reason: Information for rural and urban populations for countries could be placed side by side allows comparisons to be made (1) can be enhanced by colour (1), 2 graphs could be drawn one for rural and one for urban so differences can be clearly noted (1).

Technique: Divided Bar Graphs (1)
Reason: A clear way of showing data in percentages (1). Graphs should be placed one above the other to allow comparisons between countries to be made (1).

Technique: Table (1)
Reason: Actual % can be shown allowing information to be ranked for each country (1). Differences can be clearly seen (1).

Accept any other appropriate technique and reason.   **4 ES**

9. *1 mark for each valid point. 2 marks for a developed point.*

Less goods will be sold (1). Less money coming into the country (1). Unemployment may rise (1). It will affect its rate of development (1). Can not compete fairly with countries which don't have to pay (1).   **3 KU**

10. *Either/or acceptable.*
1 mark per valid point, 2 marks for a developed statement.

Most likely answer would be:

Short Term Aid
So many people were made homeless that they would need to have emergency shelters (1). Water supplies would have been contaminated so clean water would be essential (1) to help stop further death and disease (1). People would need food because their crops would have been destroyed (1) and medicines would be required for people made ill by dirty water or by starvation (1).

Accept any other valid point and valid reasons for Long Term Aid.

Long Term Aid
Money can be used to rebuild infrastructure (1) hospitals (1) houses (1) sewage, water supply (1). New industry (1) to recreate jobs (1). Economy will have been ruined, creating unemployment (1). Money is needed to rebuild economy (1). Services such as schools need to be replaced (1).   **4 ES**

GEOGRAPHY CREDIT 2010

1. (*a*) *One mark per valid statement, two for a developed point. Mark 4:1; 3:2; 2:3 or 1:4. Maximum of 1 for grid reference.*

Advantages include easy access to the start of it – B road and railway station (1); other places of interest close to the path eg Deep Sea World and Carlingnose Nature Reserve (2); variety of coastal scenery in square 1381 (1); with potential wildlife habitats eg mudflats in square 1382 (1); features of historic interest eg cultivation terraces (125824) (2).

Disadvantages include unsightly disused quarries and very large Cruicks Quarry (1381) (1); part of the path is along a road and beside a railway in 1282 (1); industrial sites including factory and railway sidings (130819) alongside the route (1).

Or any other valid point.   **5 ES**

(*b*) *Mark 3:2 or 2:3.*

Techniques and reasons could include:

Count pedestrians/traffic entering the park at different times of the day/week/year (1): this would allow patterns of use to be identified (1); and would be first-hand up-to-date information (1).

Interview people who use the park (1): this would allow a sphere of influence to be identified (1); and also what attracts people to the park (1).

Accept any other relevant technique or reason.   **5 ES**

(*c*) *No mark for choice. Maximum 1 mark for grid references.*

If service centre chosen:
There are many schools in Dunfermline (eg 1187) (2); and also various hospitals (eg 112887) (1). Dunfermline has a variety of other services which are typical of a service centre including a college (1), police station (1) and a library (1). There are also bus and railway stations (1) and there is a football stadium at 103879 (1).

Tourist centre:
Any reference to parks (1), Fife Leisure Park (1), reference to museums (1), abbey (1), loch with sailing (1), golf course at 1185 (2).

Or any other valid point.   **4 ES**

(*d*) *One mark per valid point, two for a developed statement. Negative points about the other area acceptable. No marks for grid references.*

Answers may include:

Area X is better because: the land is more gently sloping than Area Y so it will be easier to build (1); it is not next to an A class road so there will be less traffic and it will be quieter (1); it is close to Fife Leisure Park and Halbeath Retail Park so it will be convenient for families living in the new houses (1); there are three schools within about 1 kilometre which will suit families with school aged children (2); it is closer to the motorway (M90) than Area Y, but not too close, so it is easier to get to other places such as Edinburgh (2): there are old mines and air shafts in Area Y so it may be unstable land and it might be dangerous for families living there if children get into the old shafts (2); stream and small loch in Area X – pleasant environment (1).

Area Y is better because: it is closer to the centre of Dunfermline so it is easier for residents to get to the shops and to get to jobs in the CBD (2); it is next to an A class road so there is better access (1); it is closer to leisure facilities such as off road cycle route and footpath (1); Area X has an electricity transmission line running through it which will look ugly (1).

Or any other valid point. **5 ES**

(e) *One mark per valid point, two for a developed statement. Yes/no answers acceptable. No marks for grid references.*

Answers may include:

Yes: it has excellent road access from a B class road which links within less than 1 kilometre to a motorway, so it will be easy to transport goods in and out (1); there is a large market close by in Dunfermline (1), from where it will also be easy to get labour (1); most of the land is not too steep so it will be possible to use machines to grow crops (1), gently sloping south facing land will be warm, sunny and is likely to be well drained, making it ideal for growing crops (2).

No: it is so close to an urban area that there may be problems with vandalism and people damaging crops by walking through the fields (2); there are air shafts just to the north of the farm which could cause problems for machines working in the fields (1); being so close to Dunfermline the farm's land will be under threat from urban expansion (1); the steep hill indicated on the B road shows that some of the land may be too steep for machinery, so the farm will be limited in what it can do with this land (1); heavy traffic related to retail park may cause inconvenience for farmer (1).

Or any other valid point. **4 ES**

(f) *One mark per valid point, two for a developed statement. No marks for grid references.*

Answers may include:

There is lots of flat land making it easy to build on and there is space available for the industries to expand on to (2); there are excellent transport links with railway sidings and the docks at Rosyth making the estate very accessible (2); goods can be transported by rail or sea helping to reduce the number of HGVs on the roads (1); there is a large pool of labour nearby in Rosyth, North Queensferry and Dunfermline (1); road links are also good as there is access to the M90 motorway close by (1); HM Naval Base nearby may be a customer for some firms (1).

Or any other valid point. **4 KU**

2. *One mark per valid point, two marks for an extended point. Both sides of the meander must be referred to for full marks. Maximum of 1 mark for correct identification of location of process.*

At point P the river flows faster, as this is the outside of the meander (1); the river has a lot of energy at point P, so erodes the bottom of the river bank (1); as well as the water hitting the banks, pieces of sediment may also be thrown against the river banks wearing them away (1); the river undercuts the bank creating a steep slope (river cliff) (1) and deepening the river bed (1).

At point Q the river flows slower as this is the inside of the meander (1); the river has less energy on the inner bank causing deposition to take place (1); making the river shallow (1); and creating a river beach (1).

Or any other valid point. **4 KU**

3. *For full marks both similarities and differences must be mentioned.*

Similarities
In both summer and winter, wind speeds are light or calm towards the centre of the anticyclone (1); in both summer and winter winds circulate in a clockwise direction (1); there are few clouds in the sky at both times of year (1); anticyclones bring a spell of settled weather in both summer and winter (1); there may be mist in the morning at both times of year (1).

Differences
Anticyclones bring hot weather in summer but very cold weather in winter (1); heat waves can occur in summer if an anticyclone remains over the UK for a period of time (1); however during winter severe frosts are common at night and during the day (1); cooling of the ground leads to morning mist in summer, but in winter fog lasting part/all of the day is common (1); warm moist air rising from the ground can form thunderstorms in summer (1).

Or any other valid point. **4 KU**

4. *One mark per valid point, two for a developed statement. Yes/no answers acceptable. No credit for simple lifts.*

Population increase puts pressure on the land (1); trees cut down and vegetation removed to make way for housing leaving land unprotected (2); trees cut down to provide more farmland as increasing population requires more food (2); less fallow land as more food is required to feed the people (1); more trees are needed for shelter and firewood, so soil exposed to weathering and erosion (2); overcultivation occurs and the soil loses fertility (1); and the desert spreads (1); also some years are wetter/drier than others, causing drought (1) and vegetation will die (1); there are no roots to hold/bind the soil (1) so it can be blown by the wind (1) or eroded by the people/animals trampling over it (1).

Or any other valid point. **5 ES**

5. *Mark 2:4; 3:3 or 4:2. One mark per valid point, two for a developed statement. Maximum 3 marks for discussion of advantages/disadvantages of location. Answer must refer to advantages/disadvantages for full marks.*

Advantages
Provides jobs for local people (1); many wealthy people in neighbouring Aberdeen who could afford to use it (1); good accessibility from the airport, city bypass and A-class roads (2); it will boost the local economy (1); sand dunes provide an excellent landscape for developing a golf course (1).

Disadvantages
People living in Balmedie will be disturbed by increased traffic (1); it will ruin the habitat of endangered birds (1), and despoil a Site of Special Scientific Interest (SSSI) (1); it's not needed because there are already three golf courses in the area (1); the many new buildings will be a visual eyesore and spoil the view from the nearby Country Park (2); since there is low unemployment in Aberdeen it may be difficult to get enough people to work on the development (1).

Or any other valid point. **6 ES**

6. *Candidate may refer to features, age, appearance or function of the zones. Mark 4:2; 3:3 or 2:4.*

Answers may include:

CBD: will have tall buildings because space is scarce (1) and land values are high (1); these buildings are likely to be occupied by high order services such as department stores, which need a central location to attract large numbers of customers (2).

Inner city: will have old factories because this is where industry set up first (1); close to old housing because people had to walk to work (1); the housing will be tenements or terraces which permit high housing densities (1); laid out in a grid-iron pattern to save space (1) and because little thought was given to environmental quality (1). The grid-iron pattern will encourage through traffic causing air and noise pollution (1).

Outskirts: Modern housing has been built away from the unpleasant conditions of the inner city for workers and managers who could afford transport (2); and may be detached or semi-detached houses with gardens for a pleasant environment (1). These houses usually have driveways and garages as many people now commute to work by car (1) and there is more space to build them as the land here is cheaper (1).

Green Belt: this land is protected by strong planning regulations to prevent urban sprawl from happening (1) which would use up more and more countryside and farmland (1). Recreational amenities may be found in the green belt such as golf courses/country parks, provided for the nearby population of the city (2); sometimes bypasses or ring roads are built here to ease traffic flow around the city (1).

Or any other valid point. **6 KU**

7. *No marks for straight lifts. One mark per valid point, two for a developed statement.*

Answers may include:

Redland Farm has diversified (1) and now earns a large percentage of its income from new activities such as rented holiday cottages because its old farming activities were not profitable enough (2) and new sources of income help to keep farm viable (1); some crops have become less fashionable, such as oil seed rape (1) while others, such as organic produce bring in a good income as they are becoming more popular, because people want to eat healthy food (2); also government and EU grants are now being given for different crops and activities compared with 1980 (1) such as grants for protecting the environment instead of growing crops (1). As tourism has grown, Redland Farm can now earn a good income from tourist related activities such as quad biking (1).

Or any other valid point. **6 KU**

8. *One mark per valid point, two for a developed statement. Maximum 1 mark for simple links eg dry areas are sparsely populated.*

Northern areas of Kenya are sparsely populated because these areas receive very little rainfall so there is a lack of water for domestic use and to irrigate crops (2); the south west of Kenya is wetter so population density is higher here as crops and livestock farming can take place (1); population density is high in and around Nairobi city as there are plantations nearby and these require many workers (1); the major industrial areas have a high population density as there are job opportunities for the people and services like schools, hospitals and shops (2); along the south eastern coast the population density is quite high as these areas have more rainfall and more pleasant to live in (1); Mombasa is a port so there are jobs here (1); this area also attracts tourists so there are lots of jobs available (1).

Or any other valid point. **4 ES**

9. *One mark per valid point, two for a developed statement. Yes/no answers acceptable. Or any other valid point.*

Yes: India will have to grow or import more food as there are more mouths to feed (1); there will be a large number of young people who will grow up and want to have their own families, so the population will continue to grow

(1), putting even more pressure on resources like housing and health care (1); there might be a shortage of schools and trained teachers as the country will find it difficult to pay for them (1); this will result in a lack of knowledge about family planning and birth control (1); living standards are likely to drop and poverty and disease will increase as people cannot afford to feed themselves (2) or pay for hospital/medical treatment (1).

No: India's growing population will provide a large active workforce (1); which will be able to produce more food (1) as well as goods for export which will help India's economy to grow and help to improve its trade balance (2); there will be large number of fit people to join India's armed forces to help defend the country (1).

5 ES

10. (a) *One mark per valid point, two for a developed statement. Maximum one mark for quoting export figures.*

Overdependence on raw materials; for example Zambia depends on copper for 87% of its exports (1); or, less income if price drops (1); less to buy imports (1); have to borrow money (1); increased debt (1).

Or any other valid point. **4 KU**

(b) *Mark as 2:3 or 3:2 for choice of techniques/ justifications. The same reason can be credited only once.*

Pie charts (1):	these would be suitable as they are already in percentages (1); a series of charts could be drawn, one for each country; this will make comparison easy (1); the different segments could be highlighted using colour (1).
Divided bar graphs/: Bar Graphs (1).	a divided bar graph could be drawn for each country and could be placed in rank order (1); this will make it easy to see at a glance which countries depended the most on just one or two commodities (1). A bar graph would also show this in the same way, with the highest percentage shown first (1).
Tabulating (1):	a table could be used to put countries in rank order, the most dependent countries appearing first (1); the table could be subdivided into categories to show countries in groups (1); such as over 90%, between 80-90% (1); these can also be colour coded to highlight these differences (1).

Or any other valid technique and justification. **5 ES**

11. *One mark per valid point, two for a developed statement.*

Answers may include:

They are often more successful because aid is focused on small scale projects helping those in greatest need (1), which usually have little or no effect on the environment (1), and don't try to change the local way of life (1); encourages local people to work together and help themselves (1); developing their own knowledge and skills (1); promoting the use of appropriate technology (1) which is cheap and easy to repair (1); they sometimes work as pressure groups and therefore are able to influence decision-making (1); for example Oxfam and other charities work together to campaign for debt relief for the poorest countries in the world (1); they help to focus aid given by governments, individuals and businesses (1); they raise public awareness of global issues (1) eg Beat Poverty campaign by Save The Children raises awareness of and fights child poverty (1).

Or any other valid point.   **4 KU**

1. (a) *1 mark per valid statement. Maximum 1 mark for grid reference. References to valley acceptable. Appropriate 4 figure grid references are acceptable.*

The River Wear is fairly wide (1); it is flowing from South to North (1); it is meandering (1), for example at 273418 (1); it is embanked (1); it is joined by a tributary at 286421 (1); it is on a gentle gradient (1); in its middle/lower course (1); there is evidence of deposition on the inside of the bend at 277431 (1).

Or any other valid point.   **3 KU**

(b) *1 mark per valid statement, 2 for a developed point. No marks for grid references.*

Area A is almost surrounded by a meander (1), so is good for defence (1); the castle at the meander neck supports this (1); there is flat land in the middle of Area A to build on (1); water is available from the river (1); there are trees, so wood available for fuel/building (1); flat land nearby for farming/food supply (1). Bridging point (1).

Or any other valid point.   **3 KU**

(c) *1 mark per valid statement, 2 for a developed point. Answers which agree/disagree acceptable. No marks for grid references.*

Much of Area B is steeply sloping, and would be difficult to build on (1); there is marshland which would have to be drained (1); this would be expensive (1); the small lakes may be dangerous for children if homes were built here (1); there could be many objections, including from farmers who use the land (1); from groups who use the pathways/bridleways (1), or from people who think this is a good area for wildlife (1); like the RSPB (1).

Or any other valid point.   **4 ES**

(d) *1 mark per valid statement, 2 for a developed point. No marks for grid references.*

It is close to Durham where there are plenty of job opportunities (2); commuters can drive into the city easily on the A177 (1); High Shincliffe is mostly residential (1); with modern housing estates (1); there are few services (1); and these are low order, eg public house (1); there is no industry (1) so people would need to commute to get jobs (1); it is a pleasant environment surrounded by countryside (1).

Or any other valid point.   **3 ES**

(e) *1 mark per valid point, 2 for a developed point. Yes/no answers acceptable. No marks for grid references.*

Yes: It is on the outskirts of Durham so land will be cheaper (1); the land is flat/gently sloping so it is easy to build (1); main roads A167/A177 make the development easily accessible from Durham and surrounding areas (2); the surrounding colleges will supply a highly qualified work force (1); and may supply opportunities for business partnerships (1); students from nearby colleges eg Van Mildert College (1); will be able to stay in the residences (1); the land already has a golf course and will be easy to convert to parkland (1); the luxury housing will have good views of the parkland and countryside (1), and it is not far for residents to commute to Durham for work (1).

No: The golf course will be destroyed (1), reducing the recreational facilities available to the people of Durham (1); woodland may need to be chopped down to make room for buildings (1), which will result in a loss of habitat for wildlife (1), and may spoil the scenery (1); the development is hemmed in by roads and buildings, so there is no room for expansion (1); there will be increased traffic on the A167/A177, which may result in traffic congestion (1), and higher levels of noise/air pollution (1).

Or any other valid point. **4 ES**

2. *1 mark per valid point, two for a developed point. No marks for choice. Negative points about the other sketch acceptable.*

Correct choice is A (1).

The river is narrow (1); and it is flowing through a v-shaped valley (1); the valley sides are quite steep (1); the boulders in the river are large (1); there is little evidence of human activity (1); in Sketch B the river is meandering (1); which is more typical of the middle/lower course (1); the valley sides are flat so could not be the upper course (1); the river valley is much flatter in the middle lower courses (1).

Or any other valid point. **4 KU**

3. *1 mark per valid point, two for a developed point. Maximum 1 mark for a straight lift. Yes/no answers acceptable.*

Yes: The snow makes roads difficult to travel on (1), causing them to be dangerous (1), and makes accidents more likely (1); travelling time would be longer making many people late for work (1); airports close causing disruption for air travellers (1); the poor weather conditions mean that many people do not go to work (1), so less money is made by shops, factories, businesses (1); more child care needed as schools are closed (1); pavement will be icy resulting in more accidental falls (1).

No: Children are happy to get an extra day off school (1); snowfall means ski resorts can remain open (1), increasing business, and making them more money (1); people can buy suitable clothing for cold weather, eg gloves, scarf's, increasing sales of these items (1); more salt and shovels are sold, as slippy paths and driveways needed to be cleared, increasing shop profits (1).

Or any other valid point. **4 ES**

4. (*a*) *1 mark for each correct station.*

Climate Graph	Climate Station
A	2
B	1
C	4

3 KU

(*b*) *1 mark per valid point, 2 for a developed point.*

Graph A: People traditionally travelled from place to place in search of food and water (1), because lack of rainfall makes farming difficult (1); dams have been built across rivers to provide water (1); irrigation is needed to grow crops (1) which makes farming/food expensive (1), although cash crops can be grown for export (1); water may need to be obtained from the sea using desalination plants (1); high temperatures may cause heatstroke (1).

Graph B: High temperatures and heavy rainfall makes rainforests inhospitable therefore few people live in there (1) heavy rainfall washes nutrients out of the soil/leads to soil erosion, therefore land is farmed by shifting cultivation (2) or other forms of sustainable agriculture, eg agro-forestry (1); heavy rainfall causes flooding so houses built on stilts (1); mosquitoes flourish in this climate, causing malaria (2).

Graph C: Low temperatures mean the ground is permanently frozen (permafrost) which makes building difficult (1); houses need to be built on stilts so heat from buildings doesn't thaw the permafrost (1) and make buildings subside (1); cars have to be plugged into electricity supply overnight to stop engines freezing (1), and even petrol can freeze (1). Too cold for farming (1), so people traditionally obtained food by hunting (1), so their diet is mostly made up of meat (1); other produce and food must be flown in (1), which is very expensive (1); residents in danger of frostbite if they don't wrap up in winter (1).

Or any other valid point. **3 KU**

5. (*a*) *1 mark for each correctly placed line. 1 mark for labelling.*

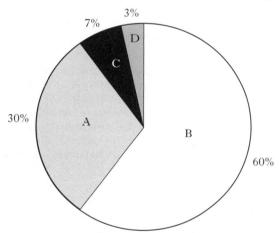

Letter or name or percentage of activity acceptable. **3 KU**

(*b*) *1 mark per valid point, 2 for a developed point.*

eg Logging:
The habitat of animals is destroyed (1), which could lead to their extinction (1); the native people lose their food supply as the removal of trees means less animals to hunt (1); and traditional medicines may be lost (1); the nutrient cycle is broken and land becomes infertile (1); soil erosion occurs as there are no tree roots to hold the soil in place (1); soil is washed into rivers leading to flooding downstream (1); large logging machinery causes air/noise pollution (1), and dragging trees out of the forest destroys the natural vegetation (1); outsiders bring disease, which the native people have no immunity against (1).

eg Cattle Ranching
Trees are destroyed to make way for cattle ranching (1); animals and people are forced off the land (1); the land becomes overgrazed and infertile (1), the forest does not regenerate and soil erosion occurs (2), the cattle ranchers move on and destroy even more rainforest (1).

Or any other valid point. **3 KU**

6. *1 mark per valid statement, 2 for a developed point. Yes/no answers acceptable.*

Yes: Playing fields replaced by supermarket takes away a valuable leisure facility (1), and may lead to increased crime (1); supermarket destroys the open aspect of the playing fields (1), and will lead to increased noise and air pollution (1); community spirit in tenements will be lost (1), as people are forced to move out (1); only wealthy people can afford the expensive apartments (1); using church buildings as pubs and night clubs might paint a negative view of the church (1).

No: A lot of new jobs will be created (1), during construction and once the developments are completed (1), lowering the unemployment rate (1) and boosting the local economy; the old factory eyesore will be removed (1), improving the appearance of the area (1).

Or any other valid point. **4 ES**

7. *1 mark per valid point, 2 for a developed point. Yes/no answers acceptable.*

Yes: It is a good place for farming because it has some gentle sloping land for growing crops (1), and also some hills which could be used for sheep (1); the narrow strips of woodland will provide shelter for livestock, especially at lambing time (2); the farm could make money by offering walkers bed and breakfast (1); Dufftown is not far away and could provide services and extra workers if needed (2); the farm is quite accessible as there is a B class road (1).

No: The farm is quite high up and so the weather will be harsh (1), this will make it difficult to grow crops and might make it hard for lambs to survive in the spring (2); walkers on the long distance footpath might create problems such as leaving gates open (1), or dropping litter which could harm livestock (1); it is in a remote areas so taking goods to market might cost more (1).

Or any other valid point. **4 ES**

8. (*a*) *Mark 2:2 or 3:1 For full marks at least two techniques must be given.*

Technique	Reason
Interview former mill workers (1).	They would be able to tell you how the closure has affected them (1), and they might be able to give the pupils information on how many workers had got new jobs (1).
Give questionnaires to local businesses (1).	This would give the pupils an idea of how the closure of the mill had affected local shops and restaurants (1), for example whether takings and custom had dropped as a result (1).
Fieldsketch/ photograph the site (1).	Could be compared with photographs of the mill when it was working (1) to see if the site has become run down or improved (1).

Or any other valid point. **4 ES**

(*b*) *1 mark per valid point, 2 for a developed point. Advantages must be explained for full marks. Maximum 1 mark for description.*

It is on quite flat land, so it is easy to build on (1); there is a water supply next to the site which could be used in manufacturing (1); it is a very accessible site next to the A96, giving good transport links for imports and exports (2); Aberdeen is only 15 miles away and could provide a big local market for produce (1); Aberdeen airport is close, making it easy for people to fly in and out on business (1); there is a local workforce in Inverurie, but workers could even travel from Aberdeen (1).

Or any other valid point. **4 ES**

9. (*a*) *1 mark for a valid point, 2 for a developed point. At least two age groups must be described for full marks. Maximum 2 marks for one age group.*

The population in the 0-19 age group is projected to decrease (1), from approximately 11 million to 9 million (1).
The population in the 20-59 age group is projected to decrease (1), from approximately 33 million to 29 million (1).
The population in the 60+ age group is projected to increase (1), from approximately 14 million to 18 million (1).

3 ES

(*b*) *1 mark for a valid point, 2 for a developed point. NB Answers must identify links between the two maps. No explanation required.*

There is a low population density where there are mountains (1), and marshland (1); population densities are moderate to high along the river Po/Po valley (1), and the lowland areas around the coast (1), particularly where coastal resorts are located (1); population densities are high near industrial towns (1) eg Milan/Naples (1), and cultural and historical centres (1) eg Rome/Florence (1).

Or any other valid point. **4 ES**

10. (*a*) Developing countries export mostly primary goods, which are cheap (1), so don't earn a lot of money (1), and their price can fall as well as rise (1), causing hardship for producers who depend on them (1); developing countries don't export many manufactured goods, which are more profitable (1); this often means they can't afford to buy things they need without loans (2), so debt and poverty increases (1).

Or any other valid point. **3 KU**

(*b*) *Mark two for techniques, two for reasons. Two reasons can be given for the same technique. There is no additional credit for giving the same reason twice.*

Technique	Reason
Divided bar graphs (1).	These are good for showing percentages/share of a total (1), and are easier to produce than pie charts as there is no need to calculate angles/draw circles (1).
Bar graph(s) (1).	Columns can be placed side-by-side for easy comparison (1), and can be coloured to emphasise differences/enhance presentation (1).
Table (1).	Allows information to be read without having to refer to a key (1).

Or any other valid point. **4 E**

11. *1 mark per valid point, 2 for a developed point.*

Low cost schemes mean no need to borrow money (1) and get into debt (1); simple tools and equipment can be maintained by local people (1), so scheme is sustainable (1); employing local people creates jobs (1), and increases income/reduces poverty (1); rural schemes help people stay in their own area (1), avoiding family break-ups (1), and prevent the need to move to overcrowded/squalid shanty towns in cities (1).

Or any other valid point.   **3 KU**

GEOGRAPHY CREDIT 2011

1. (a) *1 mark for each correct answer.*

hanging valley	435663
truncated spur	476683
corrie	467677

3 KUa

(b) *1 mark for a single point. 2 marks for a developed point. For full marks reference must be made to at least two of the land uses. Maximum 1 mark for a grid reference.*

Answers could include:

Deer stalking:
On the highest land where it is too cold and soil is too thin even for trees to grow (2). Deer are nimble so can cope with steep slopes and can survive on rough grazing (2).

Forestry:
Most of this land is above 250m where it is too cold for crops to grow (1) and the growing season is too short (1). Soils are acidic and rainfall is high, but coniferous trees can grow in these conditions (1). Many of the slopes are too steep to use machinery (1).

Mixed farming/settlement:
Land is lower and climate warmer, so more suited to settlement and cultivation, producing a pattern of scattered farm houses (2). Arable farming can take place on the flat alluvial soils of the Peffery flood plain and livestock can be grazed on the steeper sloping land (2).

Ancient fort:
Good defensive site on top of a hill (1), from which one could observe advance of attackers and more easily repel them (1).

Or any other valid point.   **5 KUb**

(c) *1 mark for a single point. 2 marks for a developed point. Both advantages and disadvantages must be mentioned for full marks.*

Answers might include:

Advantages:
South-facing slope therefore skiers will be in sunshine (1). In a corrie into which snow will be blown by the wind (1). Slopes vary in steepness which will cater for skiers of all ability levels (1). Most of the area has no cliffs so is safe (1). Area is above 700m which will be cold enough for snow to lie (1).

Disadvantage:
South facing, so snow will melt in the sun and not lie long (1). There are cliffs in the NE which will be dangerous (1). It is inaccessible (1), the nearest road being 5km away (1).

Or any other valid point.   **5 ESc**

(d) *1 mark for a single point. 2 marks for a developed point. Maximum 1 mark for grid references.*

Answers may include:

Agree
There is a castle at 485593, a symbol stone, hut circles, field systems and a chambered cairn (2). Tourist facilities and points of interest are the maze and museums (1). Recreational activities include the cycle trail and golf course (1). Hill walking can be done on Ben Wyvis (1). There is a variety of habitat due to the range of altitude from sea level to over 1000m (1) with farmland, marsh, forest, moorland and high mountains (2). The landscape is very scenic with mountains, rivers and lochs (1). There are so few roads that most of the area is unspoilt (1).

Disagree
There is a lack of historic features (1). A large area is blanket planted with coniferous trees with straight edges to the plantations which looks ugly and unnatural (2). Lack of roads will make it difficult for the public to enjoy the area (1).

Or any other valid point.   **5 ESc**

(e) *1 mark per valid point. 2 marks for an extended point. No marks for grid reference. Yes/no answers acceptable.*

Answers may include:

Yes: The area is flat for parking caravans and pitching tents (1). There are pleasant views over the water (1). It has good road communications allowing easy access to the area by car (1). There is also a railway station close by for visitors arriving by train (1). There is a hospital near the site in case of accidents (1). The town of Dingwall is within walking distance for provisions (1). There are leisure activities close to the location eg leisure centre, museum and a castle (546601) (2). There are mountains close by for walking and climbing as well as many forest walks (2).

No: The caravan and camp site is close to a rifle range which could be noisy and dangerous (1). There are works at 560585 which would be an eyesore (1). There is a danger area off shore so water activities would not be possible (1).
There is marshland close by (556591) to the area could get water logged making it difficult to pitch tents (2). The railway line is close by which could be noisy and dangerous for children at play (1).

                                        4 ESb

2. *For full marks mention must be made of the formation of the arête. 1 mark for each valid point. 2 marks for a developed point. Credit will be given for relevant diagrams.*

Snow collected in hollows then turned to ice (1). This ice eroded the mountain on all sides creating corries (1). The back walls of the corries were eroded back towards each other (1) by the processes of plucking and abrasion (1) until a narrow knife-like ridge was formed between them (1). An arête was formed where two corries formed back to back (1).

                                        4 KUb

3. *Answers must refer to the synoptic chart. Answers should be explanation. Allow minimum of two marks for description of differences in weather.*

There is a warm front close to Glasgow causing cloud cover and wind (1). The isobars are close together so Glasgow will be experiencing strong winds (1) which will continue to increase as the front gets nearer (1). In the next few hours the front will be over Glasgow bringing steady rainfall (1). There are no

fronts close to London so it will be drier and clearer (1). The isobars are further apart so it will be less windy (1). The warm front is further away so it will take longer to reach London, allowing the celebrations to go ahead (1).

                                        5 KUb

4. *1 mark per valid point. 2 marks for an extended point. No marks for straight lifts from diagrams. Yes/no answers acceptable.*

Agree
Indonesia has a huge international debt and palm oil will bring in money which may help to pay off part of the debt (1). The income generated from the sale of palm oil may mean that Indonesia does not have to borrow more money and go deeper into debt (2). The development of palm oil plantations will provide jobs for many people (1). The income from palm oil is used to improve the standard of living of the local people (1). The demand for products made from palm oil is rising so Indonesia is guaranteed income in future years (1).

Disagree
More than 40 million hectares of rainforest has been lost in the last 45 years (1) and the amount of rainforest loss is still increasing (1). The rainforests are home to many animals which could become extinct (1) as logging and burning destroy their habitat (1). The native way of life of the rainforest people is threatened (1) as destroying the rainforest restricts the land available for hunting (and farming) (1). Large areas are cleared by burning which increases the amount of CO2 emissions (1) leading to increase of greenhouse gases and global warming (1).

Accept any other valid point.   **5 ESb**

5. (a) *1 mark for a valid statement, 2 marks for a developed point. Negative/positive points about the other locations acceptable.*

Location 1 – Inner City
Close to the CBD for work (1) and very close to a regional shopping centre where they could do most of their shopping (1) reducing travelling cost (1). The houses here are mostly terraced housing so could be cheaper (1). There is a secondary school nearby for their teenage children to attend (1). They are near main roads and a train station for easy travel to the countryside (1). They will have a good riverside view (1).

Location 2 – Inner Suburbs
There is a secondary school close by for their teenage children to attend (1). The environment would be nicer because of more open space (1). Houses are likely to be more modern than those in location 1 (1). There is a mixture of council and private housing to choose from (1). Some of the houses will be detached or semi-detached (1) with gardens suitable for families (1). Near main road for easy access to the CBD and the countryside (1). There is likely to be less traffic as there are fewer main roads compared to location 1 (1).

Location 3 – Outer Suburbs
Most modern private houses far from the city centre (1). Houses will have both front and back gardens (1) and garages (1). House will be mostly detached and semi detached with bay windows (1). They are well planned, spacious and safe (1). There is a regional shopping centre close by so no need to travel to the CBD for shopping (1) saving on travelling cost. Closest to the countryside for walks (1). There is a golf course close by if they fancy a round of golf (1). The air is much cleaner as there are few main roads/less air pollution (1). There is a main road nearby if they need to go to the city centre (1).

Or any other valid point.   **5 ES**

(b) *At least two techniques must be described. Maximum of three marks if no reasons are given, or if reference is made to only one technique. Mark 2:3 3:2. Do not credit the same reason twice.*

Possible answers might include:

Technique	Reason
Field sketching (1).	Can allow comparisons to be made between the environments of the three locations when side by side (1).
Taking photographs of the three locations (1).	They can be annotated to highlight particular features (1). They can show the areas in greater detail (1).
Conduct an environmental survey of each location (1).	This could show differences in the types and quality of buildings (1) the amount of dereliction, litter and open space (1) allowing comparisons to be made (1). Will provide accurate, up-to-date information (1) of facts and not just an opinion (1).
Extract information from an OS map (1).	This will show up differences in the amount of open space and land use (1). Will allow the three separate locations to be compared without the need to travel to them (1) saving time and money (1).
Do traffic count(s) (1). Pedestrian count(s) (1).	This will show how busy each location is (1) allowing easy comparisons to be made (1). Information will be accurate, first hand and up-to-date (1).

Or any other valid technique. **5 ESd**

6. *1 mark per valid statement, 2 marks for a developed point. Yes/no answers acceptable. There are no marks for straight lifts from diagrams.*

Yes: Fuel prices have gone up a lot and so it will be much dearer to run farm equipment and for farmers to import or export products (2). As there are less EU subsidies, farmers will have to work doubly hard to make up the shortfall (1) so some farmers may go bankrupt as a result (1). A rise in insect pests due to global warming may reduce yields (1) or increase farmers' costs as they have to buy more insecticides (1). Increasing tourist numbers in the countryside could cause problems for the farmer if they leave gates open and allow livestock to escape (1) or drop litter which could harm the animals (1).

No: There are lots of opportunities for farms to increase their profits eg farms could diversify and run bed and breakfast in the farmhouse or rent out old farm workers' cottages as holiday chalets (2). Extra income could be gained by offering pony trekking to increase numbers of tourists in the countryside (1). A well run farm shop would give farms the chance to bring in extra profit (1). There are extra grants available for looking after the environment, so farms might benefit by planting hedgerows and improving wildlife habitats (1). Global warming could mean that new crops could be produced that were not possible before (1).

5 ESb

7. *1 mark per valid statement, 2 for a developed point. For full marks both advantages and disadvantages must be mentioned. There are no marks for straight lifts from diagrams.*

Advantages
More flights will be able to land and take off at Heathrow (1) and this will mean that there will be a greater choice for people wishing to travel by air (1). It will provide an economic boost to the region by creating more jobs during construction (1) and once the terminal is up and running (1). The airport will be able to function better as it will have more room and existing facilities will not be as crowded (1).

Disadvantages
There will be a lot of disruption to people's lives as over 700 houses will be demolished in Sipson (1); Animal habitats will also be destroyed (1).
Further damage will be caused to the atmosphere by increased emissions from aircraft which increase the threat of global warming/climate change (2). Some green belt land around London will be built on which defeats the whole purpose of having green belt (1).

Or any other valid point. **5 ESc**

8. *1 mark per valid statement, 2 for a developed point. Answers must be fully explained.*

This is a good location because there are universities nearby with which companies could cooperate in their research (1); they will also provide highly trained workers for companies (1). Transport facilities are excellent as there isa dual carriageway and motorway giving very good accessibility for imports/exports (1). Edinburgh Airport is also nearby allowing scientists or businessmen a convenient place to travel from (1). The City of Edinburgh will provide a good pool of labour and a large local market for products helping to keep transport costs down (2).

Or any other valid point. **5 KUb**

9. (a) *1 mark per valid point, 2 marks for a developed point. There are no marks for descriptions.*

Reasons for the changes are:
People are moving from rural to urban areas (1) because they hope to improve their standard of living (1). In the cities there are better jobs with higher wages (1). There is education for their children (1) and better health care facilities (1) which may lead to a longer life (1). Life in the countryside is hard (1) and there are high levels of poverty (1). There is little employment outside agriculture (1) and wages are low (1). The countryside lacks services such as schools and health care (1).

Accept any other valid point. **5 KUb**

(b) *Mark 2:3 or 3:2. Do not credit the same reason twice.*

Possible techniques include:

Pie charts (1). Divided bar graph (1). Bar graph (1). Line graph (1).

Reasons

Pie charts/divided bar graphs:
Data is already in percentages and would be suited to pie charts/divided bar graphs (1). A series of pie charts/divided bar graphs would show the relative proportions of the rural/urban populations over time (1).
Colour could be used to highlight the sections for easy visual comparison (1).

Bar graph:
The information could be shown on one graph (1) with different coloured bars representing the rural and urban areas (1). The bars could be compared to see changes over time (1).

Line Graph:

Line graphs are good for showing changes over time (1) and allow trends to be easily identified (1). Different lines for rural and urban areas could be used to make them easy to compare (1).

5 ESe

10. *1 mark per valid point, 2 marks for a developed point.*

The number of births per woman is low where GDP per capita is high because:

Women have access to family planning and contraception (1). Women are better educated and able to follow careers (1) this means they marry later and delay children until they are older (1) therefore having smaller families (1). Material aspirations means women work instead of staying at home to look after children (1). Infant mortality rates are low therefore less need to have so many children (1).

The number of births per woman is high where GDP per capita is low because:

There is a lack of birth control and family planning (1). Poverty/lack of health care means many children die in infancy (1) so parents have more in the hope that a few will survive (1). In poor countries children are viewed as part of the labour force and earn money for families (1). A lack of pensions and social services means children are needed to provide for their parents in old age (1). Religious or social pressure encourages people to have more children (1). Women marry young and have larger families (1).

Or any other valid point. **5 KUb**

11. *1 mark per valid point, 2 marks for a developed point. Yes/no answers acceptable. There are no marks for straight lifts from diagrams.*

Yes: British firms have a larger potential market in the EU (1), 490 million instead of just 60 million (1), and there are no trade barriers, so firms can operate easily all over Europe (1). Grants and loans from the EU can help businesses (1), especially in poor areas or areas of industrial decline (1). People from the UK can move anywhere in Europe, and so have greater job opportunities or more choice of housing area or educational provision (2). British industry is likely to be more profitable with protection from external competition (1) and the large market puts it on a more even footing with competitors in other big economies (1) like Japan and the USA (1).

No: Resentment about foreign workers taking British jobs (1). There are language and cultural difficulties (1). The UK contributes more than it gets in assistance (1). Some EU laws are restrictive (1). Trade now more difficult with commonwealth countries (1). Dearer prices (1).

4 ESb

1. (*a*) (i) *3 marks for four correct. 2 marks for three or two correct. 1 mark for one correct.*

Correct grid references are:

Corrie loch	7830
U-shaped valley	8731
Hanging valley	8029
Truncated spur with crags	8732

3 KU

(ii) *1 mark per valid point, 2 for a developed statement. Well annotated diagram(s) could be awarded full marks. Maximum 1 mark for list of processes with no detail.*

Answers may refer to change of a valley shape from V to U (1); erosion by ice (1); processes including plucking and abrasion (1); more detailed description of processes (1); change to valley = deeper, wider, steeper sides, flat floor (1).

3 KU

(*b*) *No marks for grid references. One mark for a valid point, two for a developed statement. Mark 3:1, 2:2 or 1:3.*

Advantages include: flat land/gently sloping land, so easy to grow crops (1)/use machinery (1); south-facing slope so warm/sunny (1); water available for livestock (1); close to Crieff for market (1).

Disadvantages include: damage possible if stream floods (1); land under threat from urban expansion (1); close to town so possible trespassers may cause damage to crops/nuisance to animals (2).

Or any other valid point. **4 ES**

(*c*) *No mark for choice. Max 1 mark for grid references. Lists are acceptable.*

Reason for choice could include:

Market town: meeting place of routes/valley routeways (1); services to surrounding area (1), eg hospital, a number of schools, churches (1); bridging point (1); surrounded by farms (1) hospital (1).

Tourist centre: tourist information (1); caravan site(s) (1); Hydro (1); walks (1); standing stones (1); visitor centre (1); viewpoints (1); leisure/sports centre (1); golf course (1) leisure sports centre (1).

4 ES

(*d*) *Yes/no answers are acceptable. No marks for grid references. No marks for straight lifts.*

Evidence could include:

Yes: Water available from small rivers/burns (1); no major settlements/works, so water will be clean/pure (1); farms in area so barley available (1); eg small road from distillery links quickly to main road (A85) (1); wood available for barrels from local forests (1); near Crieff for skilled labour (1); peat from high land (1).

No: Distillery only served by small road so access difficult (1); lorries from distillery may cause congestion (1) and conflict with local/tourist traffic (1); lorries and fumes from distillery may cause air pollution (1); distillery beside river, threat of flooding (1).

Or any other valid point **4 ES**

(e) *1 mark per valid point, 2 for a developed statement. Maximum 1 mark for general statements (which don't refer to map).*

Large number of visitors could cause footpath erosion on walks/trails (1); traffic congestion likely on minor road (1); large numbers of people and noise could lessen enjoyment of viewpoint (1); forestry commission land could be damaged by fires (1); only one car park, so Comrie area may be affected by inappropriate parking (1).

4 ES

(f) *1 mark per valid statement, 2 for a developed point.*

High levels of rain on hills (1); many surface streams, giving good water supply (1); as cold, cloudy conditions in high area, so little evaporation (1); U-shaped valley, so small dam creates large storage (2); land is generally poor quality so no loss of valuable/productive farmland (1) and no loss of land required for settlement/industry (1); provides water supply for Crieff (1) surface streams indicate impermeable rocks (1) so water doesn't drain away (1).

Or any other valid point. **3 KU**

2. *Yes/no answers are acceptable.*

   Yes: Outwash plain has flat land, suitable for crops (1) and for using machinery (1); outwash plain soils are well drained, suitable for cereals (1); boulder clay soils can become very fertile when drained (1); boulder clay area good for growing grass, suitable for dairy cattle (1); hilly terminal moraine good for sheep grazing (1).

   No: Steep slopes and poor soils on hilly areas (terminal moraine, drumlins) would limit farming (1); upland areas would be good for forestry (1); outwash plain can be used for quarrying which creates jobs in industry (2).

   Or any other valid point. **3 ES**

3. *1 mark per valid point, 2 for a developed statement. No mark for general description of weather.*

   There could be heavy snowfalls which would cause roads to be blocked (1) and disrupt transport services (1). Cold temperatures might cause icy pavements (1), which would be dangerous (1). Livestock could die due to extreme conditions (1).

   Accept any other valid points **3 KU**

4. (a) *1 mark for correct plotting of both temperatures.*
   1 mark for correctly joining plotted temperatures. **2 KU**

   (b) *Both temperatures and precipitation must be described for full marks.*

   The highest temperature is 26 degrees (1) in July and/or August (1), while the lowest is 12 degrees (1) in January and February (1), giving a temperature range of 14 degrees (1). Most of the rainfall occurs in the winter months (1) with the highest (111 millimetres) in December (1). There are several months in summer with very little rain (1). It is a Mediterranean climate (1).

   3 ES

5. *1 mark per valid point, 2 for a developed statement. The same reason will not be credited twice.*

   eg **Magic stones:** stone lines help to trap water when it rains (1) and stop soil from being washed away (1). This makes it easier to grow crops which help to hold the soil together (2).

   eg **Biogas:** Makes use of plentiful local resources (1) and reduces the need for gathering firewood (1), so less trees felled allowing tree roots to hold soil together (2) and give more shade to the land preventing it from drying out (2).

eg **Grazing sheep and goats fenced off:** livestock can cause overgrazing which means the grass doesn't get a chance to regrow (2); keeping them fenced in allows most vulnerable land to be protected (1) and time for grass to grow back (1); also helps to stop livestock from eating shrubs and trees (1).

4 KU

6. *1 mark per valid reason, 2 for developed points. NB: only one land use zone should be selected.*

   CBD: located where main roads meet to increase accessibility (2) and number of customers and profit that can be made (2); at bridging point for increased trade (1).

   Industrial: In industrial estates on edge of town where land is cheap (1) and there is plenty of flat land for building near the river (1); near good road links to transport raw materials and finished goods (1); workers available in nearby residential areas (1); industrial areas near CBD likely to be 19th century and built beside housing because workers had to walk to work (2); or may be brownfield developments (1); industry near river related to docks (1).

   Residential: back from river to avoid flood danger (1); avoids hills because it is difficult to build on steep slopes (1); Bridgend developed as a satellite village because it is at a bridging point (1); developed around CBD and main roads with expansion of city (1).

   4 KU

7. (a) *1 mark per valid reason. 2 marks for developed points.*

   Answers could include:

   Close to communications such as motorways and railways (1) which enable parts and materials to be brought easily to the factories and finished goods to be distributed (1). The motorways provide access for workers (1). It is close to large markets in London (1). There is a labour force nearby in towns like Aldermaston and Bracknell (1). They appear to be built on greenfield sites which are cheap (1).

   Or any other valid point. **4 ES**

   (b) *Must specify at least two techniques. The same reason will not be credited twice.*

Techniques	Reasons
Conduct traffic counts on motorways (1)	to get an impression of traffic density/air pollution (1)
Give questionnaires to local residents (1)	to find out how many local people are employed in these industries (1); data gathered in this way is easy to process (1)
Interview managers of local employment agencies (1)	to find out if unemployment has fallen since these factories were built (1)
Compare recent maps with older ones (1)	to find out how much farmland has been lost to new building
Obtain government census data (1)	to find out if population has increased (1)

   Any other valid point. **4 ES**

8. *1 mark per valid reason, 2 for a developed point. Mark 2:2, 3:1 or 1:3.*

   Advantages:
   Fields have been made bigger, creating more farmland and increasing output (2) making the farmer extra profit (1); it is easier to manoeuvre large machines around the fields (1); saves the farmer time and money looking after the hedgerows (1); removal of habitat for insect and animal pests which eat the farmers' crops (1).

Disadvantages:
Loss of habitat for birds, animals, insects and plants (1); decreasing bio-diversity in the countryside (1); removing the hedgerows can make the scenery less attractive (1); there is an increased risk of soil erosion (1) because there are no roots to bind the soil together (1) and it can be washed away by heavy rainfall (1) fewer hedges to shelter crops from high winds (1).

Or any other valid point.   **4 ES**

9. *1 mark per valid point, 2 marks for an extended point.*

   Answer should be explanation.

   Very few people live in areas that are too cold (1) as it is difficult to grow crops (1) food has to be brought in making it more expensive (1) flat land is easier to build on (1) and is easier to farm (1) mountainous areas are isolated (1) and difficult to access (1) areas with a pleasant climate are more attractive (1) areas with natural resources eg coal attract people (1) as these provide employment opportunities for people (1).

   Accept any other valid point.   **3 KU**

10. *1 mark per valid point, 2 for an extended point.*

    Children are expensive so the more children in a family the greater the financial burden (1). Women want careers so put off having children (1) or limit the number of children they have (1). Later marriages are more common which results in less children (1). Contraception/family planning clinics is widely available(1). Children are not needed to supplement the family income (1) or look after parents in old age (1).

    Or any other valid point   **3 KU**

11. *Mark 2:2. Two marks for reasons for one technique. The same reason will not be accepted twice.*

    Technique: Bar Graph(s) (1)

    Reason: Information for imports and exports could be placed side by side allowing comparisons to be made (1); can be enhanced by colour (1).

    Technique: Divided Bar Graphs (1)

    Reason: A clear way of showing data in percentages (1). Graphs should be placed one above the other to allow comparisons between imports and exports to be made (1).

    Technique: Table (1)

    Reason: Allows information to be ranked for imports and exports (1). Allows clear comparisons to be made between imports and exports as they will be side by side (1).

    Any other appropriate technique and reason is acceptable.   **4 ES**

12. *1 mark per valid point, 2 marks for a developed statement. No marks for straight lifts.*

    Long Term Aid
    Money can be used to rebuild infrastructure (1) hospitals, houses, roads, water supply (1). New industry to recreate jobs (1). Economy will have been ruined, creating unemployment (1). Money is needed to rebuild economy (1) and to restore electricity lines (1). Services such as schools need to be replaced (1). Farmers will have lost seeds and livestock (1).

    Short Term Aid
    Many people made homeless need shelter and blankets (1) and they will need fed because they will not have the means to cook for themselves (1). Helicopters would be needed to rescue people (1) who might otherwise have drowned (1). Disease can spread very quickly in these conditions and doctors/nurses would be needed to help cure people and dispense medicines (2). Water supplies would have been contaminated so clean water would be essential (1).

    4 ES

GEOGRAPHY CREDIT 2012

1. (a) *Both river and valley must be referred to for full marks. Maximum 1 for grid references.*

   Possible answers might include:

   River:
   The R Ribble flows generally in a westerly direction/towards the west (1). The river is joined by several tributaries (1) eg at 549283 (1). It meanders between 590305 and 527290/in the east (1). Its course is fairly straight between 527290 and 450277/in the west (1). The river becomes wider at 527290/towards the west (1). Several lagoons/lakes are in 4527 (1). The river becomes tidal at 552288 (1). Between 505293 and 450277 the river's banks are muddy (1). The river is in its lower course (1).

   Valley:
   The river has a wide flat floodplain (1). The valley floor is marshy eg at 4527, Hutton Marsh (1). There are embankments eg at 4828 (1). The valley sides are gently sloping (1), rising to 20m (1); although there is a steep area to the north at 568297 (1).

   Or any other valid point.   **4 KU**

   (b) *Mark 2:3 or 3:2.*
   Advantages:
   It has an accessible location with motorway junctions all round it (1). There are residential areas nearby in Walton Summit and Clayton Brook which can provide a workforce (2). The land is flat/gently sloping making it easy to build on (1). It is on the edge of Preston so the land would have been cheap (1).

   Disadvantages:
   It is difficult to expand because it is hemmed in by motorways and a railway (1). It is not an attractive area for workers with noisy motorways all around (1). Dense road network would lead to traffic congestion (1). Cause problems for delivery (1). Credit negative impact of visual/air pollution (1).

   Or any other valid point.   **5 ES**

   (c) *Yes/no answers are acceptable. No marks for grid references.*

   Yes: Varied habitats for wildlife (1) – lakes, woodland and stream (1). Reasonably quiet, since no busy roads or railways nearby (1). Land is flat for the building of the Visitor Centre (1).

   No: Noise from works could disturb wildlife (1) and effluent could pollute habitats (1). Close to residential areas so people might steal eggs or poach deer (1). If it attracts many people the minor roads could get congested (1) annoying local people (1).

   Or any other valid point.   **4 ES**

   (d) *Maximum 1 mark for grid references.*

   Bridges had to be built to cross rivers and existing main roads (2) eg 582300 (1). Crossing higher ground cuttings had to be made eg 576284 (1). On the flood plain at 580306 an embankment had to be built (1). Woodland had to be cut down at 577312 (1).

   Or any other valid point.   **4ES**

(e) *For full marks candidates must refer to at least 3 zones. No marks for grid references. Maximumimum of 3 marks for any one zone.*

Answers could include:

Dormitory Settlement
No evidence of any industry in Broughton so people must travel to work elsewhere (1). There are only low order services (1) such as the PO and school (1). There is good road access by A6 into centre of Preston (1). It is only 1km from the edge of a large town where people can work (1).

Green Belt
There is open space between Broughton and Preston which could be built on, so it is likely that development has been restricted (2).

Suburbs
The street pattern is modern geometric curves and cul-de-sacs (1). The school, leisure centre and hospital provide modern services for residents (1). It is close to the edge of town so was built more recently than other parts (1). There is more open space than in the area close to the centre (1).

Inner City
Streets are close together (1) and in a grid iron pattern (1). Large blocks by the railway nearby suggest industry (1) so this will be housing built for workers (1). County Hall (1).

CBD
There are many churches indicating the oldest part of town (1). Main roads converge here (1). It is in the centre of Preston (1). Railway and bus stations (1). Tourist information (1) and museum (1).

Or any other valid point.   **6 KU**

(f) *Mark 3:2 or 2:3. The same reason will not be accepted twice.*

Give questionnaires to (or interview) local residents (1)	*Can find out where people work (1)*
Observe and record information about job opportunities in the village (1)	*This indicates whether people need to travel for work (1)*
Map the settlement showing locations of industry/places of work (1)	
Traffic counts at morning and evening rush hours (1)	*To establish whether cars are mainly leaving Broughton in the morning and returning in the evening (1)*

                                                         5 ES

2. *For full marks all processes must be mentioned, otherwise a maximum of 4 marks are available.*

Frost shattering:
This occurs when water enters cracks in the rocks, then freezes and expands (1). Repeated over and over, this levers fragments away from the rock (1), making the back wall steeper (1). The fragments become part of the glacier's load (1).

Plucking:
This occurs where ice freezes to the rock and pulls pieces away as it moves (1). This makes the back wall steeper (1) and provides tools for further erosion (1).

Abrasion:
This occurs where rock fragments are used as tools to scrape the floor of the corrie (1), making it deeper/overdeepening it (1).

Or any other valid point.   **5 KU**

3. (a) *1 mark for each clear difference described. No marks for explanation.*

Cape Wrath has a north wind whereas Banbury has a west wind (1). It is 35 knots at Cape Wrath but calmer in Banbury at 15 knots (1). It is dry in Banbury but there are snow showers at Cape Wrath (1). There is 6 oktas cloud cover at Cape Wrath but only 2 oktas over Banbury (1). The temperature at Cape Wrath is much colder at 2C, while at Banbury it is 11C (1).

                                                         4 KU

(b) *Yes/no answers are acceptable (more likely to be no).*

Answers could include:

Yes: After the cold front has passed there will be fewer clouds and less rain (2). If they have the right gear they will be protected from the strong winds and cold temperatures (1). There could be good visibility between the showers as there will be clear air coming in from the north and this will help them find their way on the hills (2).

No: They should not go walking in the hills as there is a cold front about to arrive in the area (1) which will bring heavy rain showers (1). It will also cause the temperatures to drop close to freezing point and there could be snow (2). If they are not properly equipped they could suffer from the cold and get hypothermia (1) especially as the isobars are close together resulting in a high wind chill (1). If there is heavy snow or low cloud they could lose their way easily and need to be rescued (1) these conditions are life threatening and they should wait for a better day (1).

Or any other valid point.   **4 ES**

4. *Mark 4:2, 3:3 or 2:4.*

Advantages:
A marine national park will help to protect creatures that live in the sea (1). It will help to stop overfishing and allow fish stocks to build up again (2). Stronger laws will be introduced so that there will be less chance of marine pollution from sources such as sewage and industrial effluent (2). It might help local businesses to establish trips to see marine wildlife (1) such as the dolphin trips in the Moray Firth (1). This could boost Scotland's economy by bringing in extra tourists (1).

Disadvantage:
Activities such as trawling and oil exploration might not be allowed in marine national parks (1) and so towns and villages on the coastlines of these parks might suffer as a result (1), not just from fewer fishing and oil jobs (1) but also from the knock on effects in other industries such as fish processing and boat building and repair (1). A marine national park might impose restrictions on developments such as wave/tidal energy machines (1), but Scotland needs more renewable energy to help the environment and reduce global warming (1). Most Scottish fishermen are already fishing in ways to help preserve fish stocks so there is no need for marine national parks (1).

Or any other valid point.   **6 ES**

5. *Negative statements about the alternative site are acceptable, as are answers that refer to both sites.*

Brownfield:
Building here would rid the area of a derelict eyesore (1). Old industrial land is usually cheap to buy (1). Infrastructure such as road and power supplies would already be present (1). There would be easier access to services in the city (1) such as shopping, entertainment and schools (1). It would be difficult

to get permission to build on the greenfield site (1) which could disturb natural habitats and wildlife (1) and the developer could face protests from environmentalists (1). Building on brownfield sites helps prevent urban sprawl (1).

Greenfield:
Out of town location means land will be cheap (1) and there will be plenty of space to build (1). There is a pleasant environment which may make it easier to sell the houses (2). The land has not been built on before so will be easier to develop (1). Brownfield area would be expensive to clean up and would increase building costs (2). People may not want to buy houses on brownfield sites as there will be more noise, congestion and pollution in the city (1).

Or any other valid point. **5 ES**

6. *1 mark for valid point. 2 for a developed point.*

Most of the land is high and cold so only hardy animals like sheep can survive (1). Mostly rough grazing due to steep slopes making it difficult to use machinery (1). Thin acid soils and cold wet climate make it difficult to grow crops (2). Short growing season due to high altitude also makes it difficult to grow crops (1). Flat land in valley floor enables machinery to be used to harvest crops like barley and turnips (1) which can be used as winter fodder for livestock (1). The limited amount of lowland has deeper soil and warmer climate enabling crops to grow (2). High rainfall means grass grows better than other crops, so there is more improved pasture than arable land (2).

Or any other valid point. **6 ES**

7. *Mark 3:2 or 2:3. The same reason will only be credited once.*

Multiple line graph/3 line graphs (1).	Shows trends through time (1).
Series of divided bar graphs (1).	Shows percentages/ proportions well (1). Enables comparisons to be made between each year (1).
Triangular graph (1).	Can show all the information on one graph (1). There are 3 variables (1). Shows trends (1).
Bar graphs for each year (1). Pie charts (1).	Can enhance with colour to facilitate comparisons (1).

5 ES

8. *Answers which are all about economic **or** social effects are acceptable. Straight lifts are not acceptable.*

There will be fewer children to fill the local schools which could result in the closure of schools and staff will become unemployed (2). With fewer children some schools may be amalgamated and children forced to travel greater distances (2). In future there will not be enough young people to fill the available jobs (1) and it will be difficult to attract workers as many people do not want to live in such isolated areas (1).

Increasing numbers of pensioners means there will be less money coming into the local economy causing local services to close down (2). Providing care and services for the increasing elderly population is expensive and with fewer people of working age these services may be limited (2).

Or any other valid point. **5 KU**

9. There is free trade between member states (1). The EU provides a big market giving new members' industries increased sales and profits (2). They will not have to pay tariffs on exports to the rest of the EU (1) and are protected from competition from outside Europe (1). Their citizens will be able to look for work in other EU countries and earn higher wages (2). Countries have increased bargaining power in negotiations with other trading blocs (1). They will be able to join the Euro so people will not have to change currency when travelling in the EU (1). Poor areas/countries benefit from grants/regional aid (1). Farmers' incomes are guaranteed by CAP (1).

Or any other valid point. **4 KU**

10. *Examples will be credited. Answers could include:*

The 'middle man' is removed (1), enabling a higher/fairer price to be paid to the producers in the developing countries (1). The goods sold are often primary products which normally command a low price, so the fairer price has a significant effect (2), enabling the standard of living in these countries to improve (1).

Or any other valid point. **4 KU**

11. *1 mark per valid point, 2 marks for a developed point. Yes/no answers are acceptable. No marks for straight lifts.*

Eg if **'Agree'**:
Regular aid is guaranteed since it does not depend on donations (1). Governments can provide more money (1), so that large scale developments such as multi-purpose water schemes can be funded (2), leading to economic development in the developing country and improved living standards (2). Trade links can be forged between the donor and receiving country (1).

Eg if **'Disagree'**:
Much bilateral aid is tied aid, so the receiving country may have to agree to conditions that it does not like, such as having a military base (2) or having to spend the aid money in the donor country, which prevents them 'shopping around' for a better deal elsewhere (2). Charity aid means that no debts are incurred and there will be no interest to be paid (1) and the country's resources can be developed for its own benefit (1). With charity aid there are experienced field workers who have knowledge of successful schemes elsewhere and are committed to promoting self-sufficiency (2).

Or any other valid point. **4 ES**

Published by Bright Red Publishing Ltd, 6 Stafford Street, Edinburgh, EH3 7AU
Tel: 0131 220 5804, Fax: 0131 220 6710, enquiries: sales@brightredpublishing.co.uk,
www.brightredpublishing.co.uk

Official SQA answers to 978-1-84948-247-9
2008-2012